# **Chem**Connections
# Activity Workbook

## FIRST EDITION

**SHARON ANTHONY**

Northland College

**KEVIN L. BRAUN**

Beloit College

**HEATHER MERNITZ**

Alverno College

W. W. NORTON & COMPANY

NEW YORK · LONDON

W. W. Norton & Company has been independent since its founding in 1923, when William Warder Norton and Mary D. Herter Norton first published lectures delivered at the People's Institute, the adult education division of New York City's Cooper Union. The firm soon expanded its program beyond the Institute, publishing books by celebrated academics from America and abroad. By midcentury, the two major pillars of Norton's publishing program—trade books and college texts—were firmly established. In the 1950s, the Norton family transferred control of the company to its employees, and today—with a staff of four hundred and a comparable number of trade, college, and professional titles published each year—W. W. Norton & Company stands as the largest and oldest publishing house owned wholly by its employees.

Editor: Erik Fahlgren
Manuscript Editor: Brandy Vickers
Project Editor: Rachel Mayer
Editorial Assistant: Renee Cotton
Marketing Manager, Chemistry: Stacy Loyal
Production Manager: Eric Pier-Hocking
Permissions Manager: Megan Jackson
Text Designer: Mark Ong
Art Director: Chris Welch
Composition: Jouve
Illustration Studio: Dartmouth Publishing, Inc.
Manufacturing: Marquis Book Printing Inc.

ISBN 978-0-393-91305-7

W. W. Norton & Company, Inc., 500 Fifth Avenue, New York, NY 10110
www.wwnorton.com

W. W. Norton & Company Ltd., 15 Carlisle Street, London, W1D 3BS

5 6 7 8 9 0

# About the Authors

**SHARON ANTHONY** earned an AB degree from Bowdoin College in chemistry and mathematics and a PhD in physical chemistry from the University of Colorado. She is an associate professor of environmental science at Northland College where she teaches courses in chemistry and atmospheric science and teaches in a learning community about sustainable agriculture. Prior to arriving at Northland, Sharon was on the faculty for ten years at the Evergreen State College where she taught in learning communities about trash, the Olympic Peninsula, forensics and criminal behavior, biogeochemical cycles, climate change, and alternative energy. She coauthored the ChemConnections modules "What Should We Do About Global Warming", "Why Does the Antarctic Ozone Hole Form in the Antarctic Spring?", and "How Do We Get From Bonds to Bags, Bottles, and Backpacks?"

**KEVIN BRAUN** obtained a BS degree in anthropology and chemistry from Beloit College and a PhD in physical chemistry from the University of Arizona. After a postdoctoral position at the University of North Carolina–Chapel Hill, he joined the faculty at Beloit College where he teaches courses in instrumentation, analytical and general chemistry. His research interests include forensic applications of electrophoretic separations and chemical archaeology. In the summer, he leads faculty workshops on integrating topics in renewable energy into the general chemistry curriculum through the NSF-funded Chemistry Collaborations, Workshops and Communities of Scholars (cCWCS) program.

**HEATHER MERNITZ** has a BA in chemistry from Kalamazoo College, an MS in human nutrition, and a PhD in nutritional biochemistry and metabolism from Tufts University. Heather is an assistant professor of physical science at Alverno College in Milwaukee, Wisconsin, where she teaches courses in chemistry and biochemistry and serves as a consultant on principles and practices of assessment. She coauthored the ChemConnections module "Would You Like Fries With That? The Fuss About Fat in Our Diet" and has led numerous workshops for college faculty on using active, collaborative, student-centered learning principles in real-world contexts to teach undergraduate chemistry.

# Preface

What is the difference between a trans-fat and an omega-3 fatty acid? How is acid rain formed? How much energy is in my food? How will I use chemistry after this course? We wrote this workbook to help students discover how chemistry is used to answer these types of questions. We believe using these activities will help make the answer to the last question more obvious.

The ChemConnections Activity Workbook includes fifty-nine activities that are set in the context of societal and environmental issues. The goal is to provide instructors with pedagogically effective activities to supplement their lectures, recitation, or to use as homework assignments. Each activity was designed to take between thirty and ninety minutes, with ten of the activities appropriate for lab. These activities will help students learn chemistry content and simultaneously understand how chemistry relates to issues such as climate change, ozone depletion, recycling, greener fuels, and fats in foods.

This workbook boasts a strong contextual and educational foundation. Each activity is framed by an environmental or societal issue that is interesting to both faculty and students. We believe it is much more interesting to learn to balance chemical equations while looking at greenhouse gas reactions than by balancing chemical equations that seemingly do not apply to our everyday lives. Each worksheet begins with a central question and some basic information to set the context. Next, students are asked to work through a series of questions that ask them to apply chemical concepts to solve a problem.

To make it easier for students to apply chemical concepts to everyday life, the activities were designed with attention to pedagogy and student learning styles and use a variety of activity styles including data analysis, laboratory, worksheet, and discovery. The workbook is organized similarly to most general chemistry textbooks to allow easy incorporation into your course. Two alternate tables of contents organize the activities by topic and by theme.

Since each activity has been written as an independent worksheet, you can readily incorporate one or many activities into your existing course. The Instructor's Guide provides detailed answers to each of the questions as well as suggested discussion questions and hints about possible student misconceptions. The Instructor's Guide is available in electronic format on the Norton Resource Library.

# Acknowledgments

We would like to thank all of the contributing authors:

Howard Drossman, *Colorado College*

Tricia Ferrett, *Carleton College*

Theodore Gries, *Beloit College*

Karen Harding, *Pierce College*

Kim Kostka, *University of Wisconsin–Rock County*

George Lisensky, *Beloit College*

Kim Schatz, *University of Wisconsin–Fox Valley*

Brock Spencer, *Beloit College*

Mary Walczak, *St. Olaf College*

We would like to thank Samantha Glazier, Amy Irwin, and Sushilla Knottenbelt who served as accuracy checkers. We would also like to thank the editorial, marketing, and production team at W. W. Norton for their support and efforts in making this workbook a reality. We are especially appreciative of the efforts of Erik Fahlgren, Editor; Renee Cotton, Editorial Assistant; and Rachel Mayer, Project Editor.

Activities in the ChemConnections workbook have been drawn from or modeled after portions of the ChemConnections Modules, developed under the direction of the ChemLinks Coalition, headed by Brock Spencer, and the ModularChem Consortium, headed by Eileen Lewis, and based upon work supported by the National Science Foundation grants DUE-9455918 and DUE-9455924. Sandra Laursen served as an author and editor for the first set of ChemConnections Activities, published in the Instructor's Resource Manual for *Chemistry: The Science in Context*. We would also like to gratefully acknowledge the authors of the original modules, as well as the many faculty across the country who helped to test and refine them. Activities in this workbook have been adapted from the following ChemConnections modules, also published by W.W. Norton.

What Should We Do About Global Warming?
*Sharon Anthony, Tricia Ferrett, Jade Bender*

Build a Better CD Player: How Can You Get Blue Light from a Solid?
*Herbert Beall, Dean J. Campbell, Arthur B. Ellis, George Lisensky, Joanne Stewart*

Why Does the Ozone Hole Form?
*Sharon Anthony, Tricia Ferrett*

Would You Like Fries With That? The Fuss about Fats in Our Diet
*Sandra Laursen, Heather Mernitz*

Soil Equilibria: What Happens to Acid Rain?
*Sharon Anthony, Michael Beug, Roxanne Hulet, George Lisensky*

How Can We Reduce Air Pollution from Automobiles?
*Howard Drossman, Wayne Tikkanen, Sandra Laursen*

How Do We Get from Bonds to Bags, Bottles, and Backpacks?
*Sharon Anthony, Karen Harding*

Should We Build a Copper Mine?
*Paul Charlesworth, Mary Walczak, Doug Williams, Linda Zarzana*

# Contents

# Contents by Topic

# SOLUTIONS/DILUTIONS

# ACIDS/BASES/BUFFERS

# TITRATION

# OXIDATION/REDUCTION

# CALORIMETRY

# THERMOCHEMISTRY

# EXPERIMENTAL DESIGN/pH

# GAS LAWS

# BLACKBODY RADIATION

# SPECTROSCOPY

# VSEPR

# POLARITY, SOLUBILITY, INTERMOLECULAR FORCES

# INORGANIC CHEMISTRY/PERIODICITY

# ORGANIC CHEMISTRY

# Contents by Theme

## TECHNOLOGY & ENERGY

# Contents by Theme

## TECHNOLOGY & ENERGY

## MISCELLANEOUS

# 1 What Am I Eating?

## LEARNING GOALS

- **To perform unit conversions**
- **To practice quantitative skills such as calculating percentages**

## INTRODUCTION

Units appear everywhere in our lives, from food labels to speed limits. Understanding these units is integral to activities such as cooking, quilting, carpentry, and chemistry. The importance of keeping units straight is highlighted in the example of the NASA Mars Climate Orbiter that crashed in 1999 because of the mismatch of metric and English units in the design and construction of the spacecraft. This activity gives you examples of units that appear in the world around us and asks you to convert between different units within the English and metric systems.

### Conversion Factors for Common Unit Conversions

| | |
|---|---|
| 2.54 centimeters (cm) = 1 inch (in) | 1000 kilograms (kg) = 1 metric ton (t) |
| 1 foot (ft) = 12 inches (in) | 1 kilogram (kg) = 2.2 pounds (lb) |
| 1 meter (m) = 3.3 feet (ft) | 16 ounces (oz) = 1 pound (lb) |
| 3 feet (ft) = 1 yard (yd) | 1 hectare (ha) = 10,000 meters squared ($m^2$) |
| 16.4 centimeters cubed ($cm^3$) = 1 inch cubed ($in^3$) | 1 acre (ac) = 43,560 square feet ($ft^2$) |
| 28.3 liters (L) = 1 cubic foot ($ft^3$) | 16 fluid ounces (fl oz) = 1 pint (pt) |
| 3.79 liters (L) = 1 gallon (gal) | 2 cups (c) = 1 pint (pt) |
| 1 cubic centimeter ($cm^3$) = 1 milliliter (mL) | 29.57 milliliters (mL) = 1 fluid ounce (fl oz) |

## PROBLEMS

1. Jason is preparing curry from an Indian cookbook his British brother-in-law gave him for his birthday. Unfortunately, the British cookbook is written with units he does not typically use, so he must do some conversions before getting to work in the kitchen.

   a) The recipe calls for 450 g of diced chicken. How many pounds of chicken should he defrost?

   b) The recipe calls for 125 g of diced onions. How many 4 oz onions should he dice?

   c) The recipe calls for 100 mL of coconut milk. How many cups should he add?

2. A pound of potato chips from Hailey's favorite brand costs $4.50. A 10 oz bag of the competitor's brand costs $3.75. Which is a better bargain?

3. Brandon is traveling in France. He weighs himself at a health club because he is concerned that he has eaten too many croissants. The last time he stepped on a scale, he weighed 183 lb. The scale at the health club reads 85.0 kg. Has he gained weight?

4. Rachel loves gardening. This summer, her 10 ft × 10 ft garden space is divided equally among tomatoes, lettuce, and strawberries.

   a) The instructions for tomato fertilizer suggest applying 1500 lb/ac. How many kilograms of fertilizer should she apply to her tomatoes?

   b) A typical garden can produce four heads of lettuce per square foot. How many heads of lettuce can she expect to harvest?

   c) One of Rachel's favorite luxuries is strawberry jam made with her harvest. Each 5 pt batch of jam requires 2 lb of strawberries and way too much sugar to mention. If the average strawberry harvest is 500 kg/ac, how many pints of strawberries will she be able to make?

5. The hot dogs you ate at the barbecue last week were 75% fat-free by weight, had 275 calories, and weighed 110 g.

   a) What percentage of these hot dogs are fat?

   b) How many grams of fat does each have?

   c) What is the percentage of Calories from fat? (*Hint:* Fat has approximately 9 Cal/g.)

6. Your roommate brings home a bag of cookies, and before you know it, you've eaten the whole package. You console yourself with the fact that at least they were fat-free cookies. The serving size is two cookies and there were 22 cookies in the package.

   a) If each serving has 120 Cal and 0 g of fat, how many Calories have you just consumed?

   b) If the full-fat version of the cookies has 120 Cal and 8 g of fat per serving, how many Calories have you saved by eating the fat-free cookies?

   c) Now suppose your fat-free cookies actually have 0.48 g of fat per serving (this is legally still defined as "fat-free"). How much fat have you actually consumed eating these fat free cookies?

7. The average person in Japan consumes 65.0 kg of rice per year. Annual rice production in Japan is 8,029,000 metric tons (t).

   a) If the population of Japan is 127.9 million people, is Japan a rice importing or exporting country?

   b) How many metric tons of rice do they import or export per year?

# How Do We Convert between the Units Used in Environmental Science?

- To become comfortable converting between different units of measurement
- To use the units parts per million and parts per billion

## INTRODUCTION

A number doesn't have much value in science without a unit attached. Imagine a paramedic wheeling a patient into the hospital and telling the doctor that the patient lost "two blood" in the ambulance. Does that mean two drops of blood, or 2 L of blood? The unit makes a big difference. In this activity, you will practice converting between different units of measurement commonly used in environmental science, in both the metric and English systems.

## PART I: UNIT CONVERSION PROBLEMS

### Conversion Factors for Common Unit Conversions

| | |
|---|---|
| 2.54 centimeters (cm) = 1 inch (in) | 1000 kilograms (kg) = 1 metric ton (t) |
| 1 foot (ft) = 12 inches (in) | 1 kilogram (kg) = 2.2 pounds (lb) |
| 1 meter (m) = 3.3 feet (ft) | 16 ounces (oz) = 1 pound (lb) |
| 3 feet (ft) = 1 yard (yd) | 1 hectare (ha) = 10,000 meters squared ($m^2$) |
| 1 mile (mi) = 5280 feet (ft) | 1 acre (ac) = 43,560 square feet ($ft^2$) |
| 28.3 liters (L) = 1 cubic foot ($ft^3$) | 16 fluid ounces (fl oz) = 1 pint (pt) |
| 3.79 liters (L) = 1 gallon (gal) | 2 cups (c) = 1 pint (pt) |
| 1 cubic centimeter ($cm^3$) = 1 milliliter (mL) | 29.57 milliliters (mL) = 1 fluid ounce (fl oz) |

1. Hybrid cars are powered by both electricity and gasoline. They are becoming increasingly popular because their gas mileage is greater than that of regular cars.

   a) Assume your hybrid car gets about 38 mi/gal (mpg) of gasoline and that the current price per gallon of regular, unleaded gasoline is $2.99. The car's gasoline tank holds 11.5 gallons. How far can the car travel on a tank of gas?

   b) How many gallons of gas will be used on a 160 mi trip?

   c) What will be the cost of gasoline for the 160 mi trip?

   d) How far can the car travel on $5.00 worth of gas?

2. A sport utility vehicle (SUV) gets an average of 16.2 mpg of gasoline.

   a) Calculate how many gallons of gasoline the SUV would burn to travel 315 mi.

b) A hybrid SUV can average 30.8 mpg. How many gallons of gasoline would this hybrid SUV burn to travel 315 mi?

c) If gasoline costs $2.95 per gallon, how much money would you save driving 315 mi in the hybrid SUV compared with the original SUV?

3. It takes approximately 1200 kg of coal to produce the energy needed to make 1 kg of aluminum. If a single soda can requires approximately 15 g of aluminum, how many kilograms of coal would be needed to produce one soda can?

4. The Ford F-150 truck has an average mpg rating of 15 mpg on the highway. How many gallons would it take to drive 250 mi?

5. The world record for running a mile is 3 min, 43 s, set by Hicham El Guerrouj of Morocco on July 7, 1999. Calculate his speed in inches per second.

6. It is estimated that the United States has approximately 246,643 million metric tons of recoverable coal reserves. Convert this quantity to pounds.

7. Lake Superior is the largest of the Great Lakes, with a volume of 2900 $mi^3$. How many liters of water are contained in Lake Superior?

8. If the average American throws away 4.5 lb of trash each day and the population of the United States is estimated to be 311,000,000 people, how many metric tons of garbage does our country throw away in one year?

9. In Europe gasoline is often sold in liters. If a typical fill-up requires 11 gal of gas, how many liters are necessary?

10. A hectare (10,000 $m^2$) is a metric unit of area analogous to an acre. How many square centimeters are in a 5 ha farm?

11. Agricultural outputs are often reported in pounds per acre (lb/ac). Convert 2,000,000 lb/ac to kilograms per hectare.

12. The *Deepwater Horizon* oil spill released approximately 35,000 barrels (bbl) of crude oil per day into the Gulf of Mexico. An oil barrel is defined as 42 gal. How many gallons of oil were released each day?

13. In 2008, American farmers, primarily in the Midwest, planted 87 million acres of corn for animal feed or biofuels. How many square miles of land were used for corn production?

## PART II: PARTS PER MILLION AND PARTS PER BILLION PROBLEMS

Two concentration units you may not be familiar with are **parts per million (ppm)** and **parts per billion (ppb)**. Parts per million represents the mass (or volume, or number of molecules) of a substance per million units of mass (or volume, or number of molecules) of another substance, such as air or water or soil. The unit is referred to as ppm (by mass) or ppm (by volume) or ppm (by number of molecules). For example, the concentration of $CO_2$ in the atmosphere could be expressed as 391 ppm (by number of molecules). That would mean that there are 391 molecules of $CO_2$ per million molecules of air.

Typically, parts per million is expressed as

$$\frac{\text{mg contaminant}}{\text{kg soil}} \quad \text{or} \quad \frac{\text{mg contaminant}}{\text{L solution}}$$

Parts per billion (ppb) is defined analogously:

$$\frac{\mu\text{g contaminant}}{\text{kg soil}} \quad \text{or} \quad \frac{\mu\text{g contaminant}}{\text{L solution}}$$

Finally, to convert from parts per million to parts per billion, use the following conversion factor:

$$1000 \text{ ppb} = 1 \text{ ppm}$$

14. If there are 550 g of sodium chloride dissolved in 1 million grams of water, what are the parts per million of sodium chloride?

15. Convert 0.014 ppm to parts per billion.

16. Convert 1012 ppb to parts per million.

17. Lead is naturally present in many soils on the level of 15–40 ppm. Convert 40 ppm to parts per billion.

18. If the concentration of the herbicide atrazine, commonly used in corn production, in a water sample is 0.1 g/L, how many parts per million are present?

19. An elementary school wants to ensure the lead concentration in its soil is less than 100 ppm. If the administration sends a soil sample for analysis and the report indicates a lead concentration of 15,000 $\mu$g/g, is this soil concentration under the 100 ppm limit?

# How Does Oxygen Get to My Muscles?

## INTRODUCTION

In a process similar to the oxidation reaction that occurs within the flame of a candle, your body generates useful energy by oxidizing the chemicals that you eat (food). Just as the flame needs oxygen to sustain the reaction, the cells in your body require oxygen to produce useful energy in a sustained manner. The atmosphere on earth contains approximately 21% diatomic oxygen. How does atmospheric oxygen get to your muscle cells?

1. The average woman has 59 mL of blood per kilogram of body mass, whereas the average man has 62 mL of blood per kilogram of body mass. How many milliliters of blood do you contain?

2. The average woman has 13.9 g of hemoglobin per 100 mL of blood, whereas the average man has 15.8 g of hemoglobin per 100 mL of blood. How many grams of hemoglobin do you contain?

3. What percentage of your total body mass is hemoglobin?

4. The molecular mass of hemoglobin is 64.5 kg per mole. How many molecules of hemoglobin do you contain?

5. For nonsmokers, approximately 2.3% of the hemoglobin molecules within the blood are unable to bind and carry oxygen. As much as 6% of the hemoglobin molecules are inactive in oxygen transport in individuals who smoke. How many inactive hemoglobin molecules do you contain?

6. Each hemoglobin molecule can bind four diatomic oxygen molecules. How many grams of oxygen can each liter of your blood bind?

7. Arterial hemoglobin is approximately 95% saturated with oxygen, whereas venous hemoglobin is approximately 55% saturated with oxygen. How many grams of oxygen per liter of blood are delivered to your tissues as the blood moves through your capillaries from your arteries to your veins?

8. A normal resting heart rate is between 60 and 100 beats per minute. In physically well-trained individuals, the rate may be even lower. You can estimate your resting heart rate by placing two fingers on the thumb side of your wrist or on your neck to the side of your windpipe. When you feel your pulse, count the number of beats in 10 seconds. Multiply this number by 6 to determine how many times your heart beats in one minute. What is your current resting heart rate?

9. In an average human, one contraction of the heart will move approximately 70 mL of blood. How many grams of oxygen are your lungs currently exchanging from the atmosphere into your body per minute?

## References

Blood volume, Problem 1: Miale, J. B. *Laboratory Medicine: Hematology*, 4th ed.; C.V. Mosby Company: St. Louis, MO, 1972.

Hemoglobin levels, Problem 2: Bauer, J. D.; Ackermann, P. G.; Toro, G. *Clinical Laboratory Methods*, 8th ed.; C.V. Mosby Company: St. Louis, MO, 1974.

Smokers versus nonsmokers, Problem 5: Miale, J. B. *Laboratory Medicine: Hematology*, 4th ed.; C.V. Mosby Company: St. Louis, MO, 1972.

Written by Theodore Gries.

# Laboratory: How Can We Separate Plastics for Recycling?

## INTRODUCTION

The number of different plastics used in our society seems nearly endless. It is no surprise that the plastics used for different applications have very different physical and chemical properties. The differences in properties make it difficult to recycle mixed consumer plastics into useable products. As a result, the separation of consumer plastics prior to reuse is crucial. Any separation must be done carefully because even a small amount of contamination with the wrong type of polymer can significantly reduce the performance of the recycled material.

The presence of recycling codes on consumer plastics aids in the identification and separation of different polymers. However, the human-powered separation of comingled plastics is time-consuming and very costly. In recent years, a great deal of effort has been directed toward developing a mechanical or chemical method for separating plastics. During this laboratory exercise, you will be using physical properties to separate plastics.

## PART I: HOW DO WE SEPARATE DIFFERENT PLASTICS?

In front of you is an assortment of plastics. Imagine that you are the technical director of a plastic-recycling plant and need to separate large quantities of these plastics. Use the materials available to you to sort the plastics, but remember that hand sorting is not an economically efficient option!

Explain the chemistry involved in your sorting solution, and discuss how effectively you were able to separate the polymers.

## PART II: WHAT ARE SOME OF THE PROPERTIES OF DIFFERENT PLASTICS?

1. Examine a variety of plastic items and fill in the following table.

   - To determine whether the polymer floats or sinks in water, cut a small piece of plastic (about 1 cm × 1 cm) and put it in cold tap water.

   - Boil some water and remove it from the heat. When the bubbles stop, put your plastic sample in the water.

   - Create a saturated saltwater solution by adding some sodium chloride to a beaker of water until you have a saturated solution (some solid salt remains undissolved).

**Properties of Plastics**

| Polymer Name and Recycling Code | Is It Colored? | Does It Bend or Break? | Does It Stretch When You Pull It? | Does It Float in Cold Tap Water? | Does It Float in Hot (Almost Boiling) Water? | Does It Float in Saltwater? |
|---|---|---|---|---|---|---|
|  |  |  |  |  |  |  |
|  |  |  |  |  |  |  |
|  |  |  |  |  |  |  |
|  |  |  |  |  |  |  |
|  |  |  |  |  |  |  |

2. Two plastics that commonly are targeted for recycling from household waste are polyethylene terephthalate (PETE #1) and high-density polyethylene (HDPE #2). One of the problems of recycling such materials is separating them. Suppose you have been hired to set up a process for separating large quantities (many tons) of waste plastic that is a mixture of PETE and HDPE. Describe how you might perform this separation. Perform your separation process in the lab to confirm that it works.

Adapted by Sharon Anthony, Karen Harding, and Kim Kostka from the ChemConnections module "How Do We Get from Bonds to Bags, Bottles, and Backpacks?" by Karen Harding and Sharon Anthony.

# What Gives Fireworks Their Brilliance?

- To practice naming compounds and writing formulas

- To practice differentiating between ionic and covalent compounds

## INTRODUCTION

Nothing heightens the experience of a sporting event, festival, amusement-park visit, or holiday like fireworks. The appeal of fireworks arises from the production of light, sound, and motion. The thunderous bangs and propulsion of the burst result from rapid air expansion due to the swift release of energy, whereas the brilliant colors result from the emission of light from heated metal salts. What compounds and reactions account for the choreographed brilliance of fireworks? In this activity, you will practice naming and characterizing elements and bond types while learning about the complex reactions that occur within fireworks.

## PROBLEMS

1. A range of elements and substances are used as pyrotechnic fuels in fireworks. Classify each of the following substances found in fireworks as an alkali metal, alkaline earth metal, transition metal, metal, metalloid, or nonmetal.

   a) Antimony      b) Sodium      c) Boron      d) Strontium

   e) Iron      f) Aluminum      g) Titanium      h) Red phosphorus

2. Pyrotechnics are the result of reactions between compounds of various bond types and functions. Oxidizing agents generate the oxygen required to burn reducing agents, which generate the explosive energy required to excite the light-emitting molecules. Compounds that emit light do so by absorbing the heat energy generated by the reaction of the exothermic pyrotechnic mixture. Name the bond type(s), ionic, covalent, or both ionic and covalent, that form the following chemicals used in pyrotechnics (the purpose of each compound is denoted in parentheses).

a) $KC_4H_5O_6$
(light-emitting compound)

b) $Sb_2S_3$
(reducing agent)

c) $K_2Cr_2O_7$
(oxidizing agent)

d) Powdered sulfur
(reducing agent)

e) $SrO_2$
(oxidizing agent)

f) $BaCl_2$
(light-emitting compound)

3. The dazzling colors produced by fireworks result from light emission from excited gas-phase molecules. Heat energy absorbed by the metal atoms causes electrons in the lowest-energy state (ground state) to relocate to a high-energy state (excited state). Because of the instability of this excited state, the electron rapidly returns to the ground state and emits light, the frequency of which depends on the energy difference between the levels. Name the following metal salts:

a) $Li_2CO_3$ (red)

b) $CaSO_4$ (orange)

c) $SrCrO_4$ (red)

d) $Ba(NO_3)_2$ (green)

e) $NaCl$ (yellow)

f) $CuO$ (blue)

4. In addition to vivid colors, fireworks also use other visual features such as smoke, glitter effects, or delays in the firing train. Write the chemical formula for the following compounds used to create these features.

a)  Antimony trisulfide
    (glitter effect)

b)  Zinc oxide
    (smoke)

c)  Ammonium chloride
    (white smoke)

d)  Bismuth (III) oxide
    (crackling microstars)

e)  Barium chromate
    (delay charge)

f)  Magnesium carbonate
    (glitter-delay agent)

5. The striking colors of fireworks result from spectral emissions of excited gas-phase molecules. Metal chloride salts are ideal for this application, though their use is limited because these compounds are typically hygroscopic. Absorbed water can inhibit desired reactions or cause hazardous, unintentional side reactions. To solve this predicament, pyrotechnic designers include chlorine-donating molecules in fireworks. Predict the molecular formula for chlorine-containing compounds composed of the following:

a) Potassium and perchlorate

b) Barium and chlorate

## References

Color generation, Problem 3: www.scifun.org/CHEMWEEK/fireworks/fireworks.htm

Chlorine containing compounds, Problem 5: Wilson, E. What's That Stuff? Fireworks. *Chemical & Engineering News* **2001**, 79, p 30.

# What Are Sources and Sinks of Greenhouse Gases?

- **To learn how to translate text into chemical equations**

- **To balance chemical equations**

## INTRODUCTION

To understand why greenhouse gas concentrations are rising, you need to understand how greenhouse gases are created and destroyed. Atmospheric gases are created by sources. For instance, molecular oxygen ($O_2$) is created by photosynthesis. Conversely, sinks remove gases from the atmosphere. Respiration is a sink reaction for molecular oxygen. In this activity, you will examine sources and sinks of greenhouse gas molecules by looking at the many chemical reactions that are involved in generating and removing greenhouse gases from the atmosphere.

## PART I: WHAT HAPPENS WHEN YOU BREATHE INTO WATER? | LABORATORY EXERCISE

In this activity, you will experimentally model a set of reactions in the hydrologic cycle. As you perform this experiment, your observations will lead you to discover the sources and sinks in this system.

- Fill a large test tube half full with synthetic seawater, a solution of sodium chloride (NaCl) in water.

- Add a few drops of phenolphthalein indicator, which is pink in basic solution and colorless in acidic solution. Now add a few drops of sodium hydroxide (NaOH) solution.

1. Observe and record the results.

- Place a drinking straw into the solution and blow gently.
2. Record your observations.

- Fill a second test tube with a saturated aqueous solution of calcium hydroxide, $Ca(OH)_2$.
3. Again, place a drinking straw into the solution, blow gently, and record your observations.

4. Which greenhouse gas does this experiment involve? What is its source? What is its sink?

5. In words, write a chemical description of the sink reactions in this experiment.

## PART II: HOW DO WE BALANCE THE RELEVANT GREENHOUSE GAS SOURCE AND SINK EQUATIONS?

The following descriptions of chemical processes identify sources or sinks of greenhouse gases. Each includes a word equation; translate each of these into a chemical equation and balance the equation. When possible, designate the physical state of each substance: gas (g), liquid (l), solid (s), or aqueous (aq). You are encouraged to use your text or any other reference material you find useful.

## Carbon Dioxide

6. Respiration by animals and plants is a source of carbon dioxide ($CO_2$). During respiration, a carbohydrate (e.g., glucose, $C_6H_{12}O_6$) reacts with oxygen to produce water and carbon dioxide.

7. The combustion of fossil fuels is the main anthropogenic (human-made) source of carbon dioxide. In the United States, approximately 44% of anthropogenic emissions are from petroleum, 35% are from coal, and 19.5% are from natural gas. During combustion, a hydrocarbon (e.g., octane, $C_8H_{18}$) reacts with oxygen to produce carbon dioxide and water.

8. Cement production accounts for 1% of anthropogenic emissions of carbon dioxide. During a pyrolysis reaction, calcium carbonate ($CaCO_3$) is heated to temperatures greater than 500°C to produce calcium oxide (CaO) and carbon dioxide. CaO makes up approximately 65% of commercial cement. The other ingredients are silicon dioxide ($SiO_2$, 25%), aluminum(III) oxide ($Al_2O_3$, 4%), ferrous(III) oxide ($Fe_2O_3$, 3%), and magnesium oxide (MgO, 3%).

9. During photosynthesis, plants convert carbon dioxide and water to carbohydrates (e.g., glucose) and oxygen. Deforestation, particularly in tropical forests, contributes greatly to a reduction in photosynthesis as a carbon dioxide sink.

10. Carbon dioxide dissolves in water to yield carbonic acid, $H_2CO_3$. The oceans are major sinks for carbon dioxide. The transfer of carbon dioxide to the surface layers of the oceans occurs relatively quickly, on the order of a few years. However, it takes centuries for carbon dioxide from the atmosphere to reach the deep ocean.

## Methane

11. The decay of organic material in anaerobic environments accounts for most global methane production. During a disproportionation reaction, a carbohydrate (e.g., glucose) is simultaneously oxidized and reduced to form carbon dioxide and methane ($CH_4$). This process occurs in wetlands, rice paddies, domestic sewage, and landfills, among other locations. (A similar process called *enteric fermentation* occurs in the digestive tracts of termites and ruminants such as cattle.)

## Nitrous Oxide

12. Denitrification occurs in soils and oceans when the nitrate ion is reduced to nitrous oxide ($N_2O$) by anaerobic bacteria in the presence of water. Oxygen and the hydroxyl ion ($OH^-$) are also produced during this process. Natural sources such as oceans and soils emit approximately twice the nitrous oxide that anthropogenic sources do. Fertilizer use in agriculture and the industrial production of adipic and nitric acid are responsible for most anthropogenic emissions of nitrous oxide.

13. In the stratosphere, nitrous oxide can also react with high-energy oxygen atoms to produce nitric oxide (NO). This reaction accounts for approximately 10% of nitrous oxide destruction.

## Why Are Greenhouse Gas Concentrations Rising?

14. The atmospheric concentrations of most of the greenhouse gases are rising. What does this imply about the balance between sources and sinks? Specify with an example for carbon dioxide.

Adapted from the ChemConnections module "What Should We Do about Global Warming?" by Sharon Anthony, Tricia Ferrett, and Jade Bender.

# How Is Acid Rain Formed?

---

**LEARNING GOALS**

■ **To become familiar with the formation of acid rain**

■ **To use stoichiometric calculations to examine the acid rain formation reactions**

## INTRODUCTION

In this activity, you will look at how acid rain is formed. Acid rain is composed of sulfurous acid ($H_2SO_3$), sulfuric acid ($H_2SO_4$), nitrous acid ($HNO_2$), and nitric acid ($HNO_3$). It is formed from the gaseous products of combustion, specifically by oxides of sulfur ($SO_2$ and $SO_3$) and nitrogen ($NO$ and $NO_2$). Although some of these oxide gases come from natural sources, about 80% of sulfur oxide emissions and 60% of nitrogen oxide emissions are the result of human activities.

## SULFURIC ACID FORMATION

Most of the anthropogenic sulfur dioxide released into the atmosphere comes from the burning of sulfur-containing fuels, especially coal, which typically contains 1% to 3% sulfur in the form of pyrite ($FeS_2$). Pyrite reacts to form sulfur dioxide according to this reaction:

$$4\ FeS_2(s) + 11\ O_2(g) \rightarrow 2\ Fe_2O_3(s) + 8\ SO_2(g)$$

The $SO_2$ generated reacts in the atmosphere and eventually produces $H_2SO_3$ and $H_2SO_4$. Thus, coal-fired power plants used to generate electricity are major sources of acid rain.

## NITRIC ACID FORMATION

The formation of most nitric acid rain begins in the cylinders of vehicles powered by internal combustion engines in which the temperature and pressure are high enough to favor the formation of nitric oxide from nitrogen and oxygen gas.

Nitric oxide reacts readily with additional oxygen to form nitrogen dioxide, which then dissolves in water to form both nitrous and nitric acid. NO and $NO_2$ are frequently found together and are collectively known as $NO_x$.

$$2\,NO(g) + O_2(g) \rightarrow 2\,NO_2(g)$$
$$2\,NO_2(g) + H_2O(l) \rightarrow HNO_2(aq) + HNO_3(aq)$$

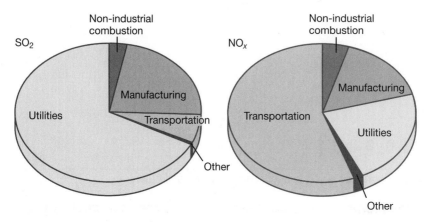

United States $SO_2$ and $NO_x$ Anthropogenic Emissions by Source, 1999
Data from http://projects.dnmi.no/~emep/areas/index.html

## CATALYTIC CONVERTERS AND SCRUBBERS

A catalytic converter, part of the automobile exhaust system, provides a surface on which much of the nitrogen oxides formed in automobile engines are reduced to nitrogen gas.

$$2\,NO(g) \rightarrow N_2(g) + O_2(g)$$

Scrubbers installed in the exhaust stream of a power plant or smelter can remove much of the $SO_2(g)$ by reacting it with calcium carbonate ($CaCO_3$) to produce a slurry of calcium sulfate ($CaSO_4$). The $CaSO_4$ is then a solid that can be disposed of in the landfill.

$$2\,CaCO_3(s) + 2\,SO_2(g) + O_2(g) \rightarrow 2\,CaSO_4(s) + 2\,CO_2(g)$$

## PROBLEMS

1. Refer to the chemical equations given in the introduction to determine the following:

   a) How many grams of $NO_2$ would be formed from 1.0 mole of NO and excess $O_2$?

   b) How many moles of $HNO_3$ would be formed from 1.0 mole of $NO_2$?

2. How many pounds of $SO_2$ are released into the atmosphere during the combustion of enough coal to fuel a 100 W lightbulb for a year? It takes approximately 715 lb of coal to fuel a 100 W lightbulb for a year. Assume that the coal burned is 2% sulfur by mass and that all of the sulfur in the coal reacts to from $SO_2$.

3. Annual emissions of sulfur dioxide in the United States are approximately 26 million metric tons.

   a) How many grams of sulfur must be in the fuel to produce 26 million metric tons of $SO_2$?

   b) Estimate how many metric tons of coal must be burned in the United States annually if 26 million metric tons of $SO_2$ is formed.

   c) The United States actually burns approximately 860 million metric tons of coal per year. What could cause your result to be different from this amount?

4. As mentioned in the introduction, installing scrubbers in power plants and smelters lowers emissions. How many kilograms of $CaCO_3$ would be needed to remove the 26 million metric tons of $SO_2$ that is produced annually in the United States?

5. Chemical treatment of fuel can remove its sulfur and thus prevent the formation of $SO_2$. Alternatively, $SO_2$ can be chemically removed ("scrubbed") from exhaust gas after the fuel is burned. Why are both of these options not available for $NO_x$?

6. Assume you smelted 1000 kg of lead(II) sulfide (PbS), and all of the sulfur was converted into $SO_2$ and released into the atmosphere.

   a) How many grams of $SO_2$ would be released into the atmosphere?

   b) Refer to the chemical equations in the introduction to determine how many grams of $CaCO_3$ would need to be used in a scrubber to react with all of the $SO_2$ formed in part a.

7. Give a concise response to the question, How is acid rain formed?

8. Give a concise response to the question, How is acid rain controlled? Be sure to address both nitric acid and sulfuric acid.

Adapted from the ChemConnections module "Soil Equilibria: What Happens to Acid Rain?" by George Lisensky, Roxanne Hulet, Michael Beug, and Sharon Anthony

ACTIVITY

# What Are Your Personal Contributions to Carbon Dioxide Emissions?

## LEARNING GOALS

- **To perform stoichiometric calculations**
- **To gain experience in experimental design**
- **To practice unit conversions, dimensional analysis, and the factor label method**
- **To employ approximation methods in problem solving**

## INTRODUCTION

The greenhouse gas carbon dioxide ($CO_2$) has a number of important sources. Balanced chemical equations allow us to *qualitatively* summarize these sources. For example, respiration and fossil fuel combustion are both sources of carbon dioxide. However, to *quantify* emissions from a specific source, stoichiometry is needed to use the quantitative information that is implicit in the balanced equation. In this activity, you will use stoichiometry to estimate: (a) how much carbon dioxide you exhale in a year, (b) how much carbon dioxide your car emits in a year, and (c) whether you would emit more carbon dioxide by flying or driving to Boston or San Francisco.

## PART I: HOW MUCH CARBON DIOXIDE DO YOU EXHALE IN A YEAR? | LABORATORY EXERCISE

Measure the amount of carbon dioxide emitted by a human in a year. You will need to:

1. Design an experiment for making this measurement with the equipment available in the lab; carry it out and repeat or refine your experiment until you have minimized your sources of error. Draw each iteration of your experimental apparatus in your laboratory notebook, and record all of your experimental information. Be sure to state your assumptions clearly, show your calculations (with units!), and acknowledge your sources of information accurately. Assume that 4% of exhaled breath is carbon dioxide.

**27**

2. Think about the uncertainty inherent in your experiment. Which step has the greatest uncertainty? How did you refine your experiment to minimize uncertainty? How could you further refine your experiment to improve your results? Answer these questions in your lab notebook.

## PART II: HOW MUCH CARBON DIOXIDE DOES YOUR CAR EMIT IN A YEAR?

Estimate the number of grams of carbon dioxide emitted by your car (or your family's vehicle) in a year. State your assumptions clearly, show your calculations (with units!), and acknowledge your sources of information accurately. You will need to:

3. Write a balanced equation for the combustion of gasoline (combustion is the addition of $O_2$ to the reactant to produce the products $CO_2$ and $H_2O$). Gasoline is a mixture of many different types of hydrocarbons. For ease of calculation, assume that it is pure octane ($C_8H_{18}$).

4. Estimate the number of grams of fuel you use in a year by converting from gallons of fuel used in a year (estimate this and be clear about your assumptions) to grams of fuel used in a year. The density of octane is 0.70 g/mL.

5. Convert grams of fuel to moles of fuel.

6. Use the balanced equation and the mole coefficients to relate moles of fuel burned to moles of $CO_2$ released.

7. Convert moles of $CO_2$ to grams of $CO_2$.

8. Compare your results from Parts I and II. Do you emit more carbon dioxide in one year by driving or by breathing? Justify your results quantitatively.

## PART III: WOULD YOU EMIT MORE CO$_2$ BY FLYING OR DRIVING TO BOSTON OR SAN FRANCISCO?

9. Estimate how many grams of carbon dioxide your car releases during a drive from [your town] to Boston or San Francisco. Gasoline is a mixture of many different types of hydrocarbons. For ease of calculation, assume that it is pure octane ($C_8H_{18}$).

10. How many grams of carbon dioxide per passenger does a Boeing 727 airplane release during the flight (from your closest major airport)? Assume that the average airspeed is 445 mi/h, the fuel consumption rate is 8000 lb/h, and the density of jet fuel is 6.7 lb/gal. The average Boeing 727 airplanes hold a maximum of 160 passengers. Although jet fuel is a mixture of many hydrocarbons, for simplicity assume that it is $C_{12}H_{26}$. The *unbalanced* reaction for the combustion of $C_{12}H_{26}$ is:

$$C_{12}H_{26}(l) + O_2(g) \rightarrow CO_2(g) + H_2O(l)$$

11. If you are concerned about contributing to rising greenhouse gas contributions, should you fly or drive? Explain your assumptions.

---

Adapted from the ChemConnections module "What Should We Do about Global Warming?" by Sharon Anthony, Tricia Ferrett, and Jade Bender.

---

# How Much Do Your Daily Activities Contribute to Greenhouse Gas Concentrations?

## LEARNING GOALS

- **To perform stoichiometric calculations**

- **To practice unit conversions, dimensional analysis, and the factor-label method**

- **To employ approximation methods in problem solving**

## INTRODUCTION

Many of your daily activities contribute to an increase in greenhouse gases. In this activity, you will perform stoichiometric calculations to estimate greenhouse gas emissions from a variety of sources. As you work through the problems, take note of the relevance of each answer to your daily life.

## PROBLEMS

For the following problems, perform stoichiometric calculations to estimate the greenhouse gas emissions for the relevant source. You are encouraged to use your text or any other reference material you find useful. However, some of the problems require you to develop skills in making estimates, and educated guesses will sometimes prove more useful than books.

1. Hummingbirds eat approximately 9.0 g of sugar water per day.

   a) If the sugar water is 25% sucrose ($C_{12}H_{22}O_{11}$), how many moles of sugar does a hummingbird consume in a day?

   b) Remember that during respiration, a carbohydrate such as sucrose reacts with the oxygen molecule to produce carbon dioxide and water. Write a balanced equation for this process.

**31**

c) How many moles of carbon dioxide are released daily by a hummingbird's respiration?

d) How many grams of carbon dioxide does a hummingbird emit daily?

e) A typical ruby-throated hummingbird weighs approximately 3.5 g. What proportion of its body weight does a ruby-throated hummingbird emit daily in carbon dioxide?

2. Propane ($C_3H_8$) is used in domestic cooking and heating.

a) A typical home in Massachusetts burns 850 gal of propane for heating over the course of a year. How many moles of propane are burned? (The density of propane is approximately 4.2 lb/gal.)

b) Propane reacts with molecular oxygen to produce carbon dioxide and water. How many moles of carbon dioxide are released per home per year?

c) How many grams of carbon dioxide are released per home over the course of a year?

d) A typical U.S. citizen emits approximately 24,000 lb of carbon dioxide by driving per year. Assuming this person lives in Massachusetts and heats with propane, how do his carbon emissions from heating compare to those from driving?

3. Trees remove carbon dioxide from the atmosphere during photosynthesis, and they actively store this carbon while they are alive. When a tree is burned, it releases this carbon again as carbon dioxide. Firewood is often sold in the unit of a cord (128 ft$^3$).

a) Assuming that one cubic foot of firewood contains 30. lb of carbon (and that all of the carbon is converted to carbon dioxide), calculate the number of grams of carbon dioxide released when a cord of firewood is burned.

b) Burning firewood is often considered to be less damaging in terms of carbon dioxide emissions than burning fossil fuels. Why do you think this is the case, when clearly burning firewood releases significant amounts of carbon dioxide?

4. Gaseous ammonia (NH$_3$) is a popular fertilizer for cornfields in the midwestern United States. An acre of corn requires roughly 150 lb of ammonia, which is applied in one allotment at the beginning of the growing season. Much of the nitrogen from the ammonia is incorporated into the structure of the plant. However, approximately 1.5% of the nitrogen from the NH$_3$ is eventually released by the plants in the form of nitrous oxide (N$_2$O).

a) How many grams of N$_2$O are released into the atmosphere for every acre of corn grown?

b) How many moles of N$_2$O are released per acre of corn?

c) How many molecules of $N_2O$ are released per acre of corn?

5. What's for dinner, beef or rice?

Methane, $CH_4$, is released from a rice paddy due to the reation given below:

$$CO_2(g) + 2\,H_2(g) \rightarrow CH_4(g) + O_2(g)$$

a) In 1 $m^2$ of rice paddy, approximately 0.125 g of $H_2$ is available each day to react with an excess of carbon dioxide. How many grams of methane ($CH_4$) are released per day by 1 $m^2$ of rice paddy?

b) Assuming that the growing season is 3 months long and that 3.5 metric tons of rice are cultivated per hectare, calculate the grams of methane released during the production of 1 kg of rice.

c) Cattle raised on feedlots in North America have an average mass of 415 kg and release approximately 37 kg of methane per head per year. If these cattle are butchered when they reach 2.0 years of age and 50% of their mass is suitable for human consumption, how many grams of methane are released during the production of 1 kg of beef?

d) Is more methane released during the production of 1 kg or rice or 1 kg of beef?

Adapted from the ChemConnections module "What Should We Do about Global Warming?" by Sharon Anthony, Tricia Ferrett, and Jade Bender.

# How Much Product Will My Reaction Yield?

- To identify limiting and excess reactants in a chemical reaction

- To use limiting reactants to determine the theoretical yield of product in a chemical reaction

- To compare actual yield with theoretical yield to determine percent yield in a chemical reaction

## INTRODUCTION

The terms *limiting reactants* (also called *limiting reagents*) and *percent yield* may sound very chemical, but you commonly encounter these concepts in your daily life. This activity will take you through some everyday examples of limiting reactants and percent yield. You will then be asked to apply these ideas to chemistry problems in real-world contexts.

## PART I: LIMITING REACTANTS

Imagine that you have volunteered to make sandwiches for a party with your friends. You buy a loaf of bread, a package of sliced turkey, a package of cheese slices, and a bottle of mustard. When you get home you open up your packages and count what you have:

18 slices of bread

10 slices of turkey

14 slices of cheese

12 ounces of mustard

1. If each sandwich has two slices of bread, one slice of turkey, one slice of cheese, and one ounce of mustard, how many complete sandwiches could be made from the preceding list of ingredients?

2. What is the limiting factor in making sandwiches? What would run out first?

3. What would be left over? How much of each ingredient would be left over?

4. In a table setting, there is one plate, one fork, one knife, and one spoon. How many table settings are possible from 10 plates, 8 forks, 12 spoons, and 10 knives? What is the limiting item?

In a chemical reaction, the reactant that limits the amount of product that can be formed is called the limiting reactant (or *limiting reagent*). The reaction will stop when all of the limiting reactant is consumed. In the sandwich example, bread was our limiting reactant. The reactant or reactants in a chemical reaction that remain when a reaction stops when the limiting reactant is completely consumed are called the excess reactant(s). The excess reactant(s) remain because there is nothing with which they can react. In the sandwich example, turkey, cheese, and mustard are available in excess. We deal with limiting reactant problems in our daily lives when we shop for food for recipes or pack outfits for a vacation. Likewise, with chemical reactions, if there is only a certain amount of one reactant available for a reaction, it doesn't matter how much of the other reactants you have: The reaction must stop when the limiting reactant is consumed regardless of whether the other reactants have been used up.

5. Consider the chemical reaction given below:

$$H^+(aq) + OH^-(aq) \rightarrow H_2O(l)$$

a) If 1 mole of $H^+(aq)$ reacts with 2 moles of $OH^-(aq)$, how many moles of $H_2O$ can be formed?

b) What is the limiting reactant?

6. Carly the chemistry student performs an experiment with vinegar (active ingredient acetic acid, $CH_3COOH$) and baking soda ($NaHCO_3$). Her professor asks her to place varying amounts of baking soda and vinegar into sealable plastic sandwich bags, quickly seal the bags, and observe the reaction. The balanced chemical equation for the reaction is given here.

$$NaHCO_3(aq) + CH_3COOH(aq) \rightarrow CO_2(g) + H_2O(l) + NaCH_3COO(aq)$$

a) Use limiting reagent calculations to fill in the following table. You will need to do your calculations on a separate sheet of paper.

| Trial | Mass of Baking Soda | Volume of Vinegar (5.0% Acetic Acid w/v) | Moles of $CO_2$ If Baking Soda Is Limiting | Moles of $CO_2$ If Vinegar Is Limiting | Limiting Reactant | Moles of $CO_2$ Formed |
|---|---|---|---|---|---|---|
| A | 0.25 g | 10 mL | | | | |
| B | 0.75 g | 8.0 mL | | | | |
| C | 1.5 g | 5.0 mL | | | | |

b) Which bag will inflate the most?

# PART II: PERCENT YIELD

7. You are making chocolate chip cookies for a party. You have followed a recipe for making the cookie batter that should produce four dozen cookies. Everything is going well until you forget to set the timer for the last batch and burn 10 of the cookies. You throw the burned cookies away and bring the rest to your party.

a) How many cookies did you expect to make from your recipe?

b) How many edible cookies were you actually able to bring to your party?

c) What percentage of the total cookies you planned on making did you actually bring to the party?

When you calculated the percentage of total cookies that you actually were able to take to your party, you were reporting a value chemists call percent yield. For a chemical reaction, the percent yield is the ratio of the actual amount that you get when you perform the reaction to the theoretical amount you calculate based on your supply of reactants.

The formula for percent yield is as follows, where actual yield is the amount of product obtained when the reaction takes place and theoretical yield is the maximum amount of product, which is calculated using the balanced equation. Yield can be reported in grams or in moles, but units must be the same for both actual yield and theoretical yield.

$$\text{percent yield} = \frac{\text{actual yield}}{\text{theoretical yield}} \times 100\%$$

8. In the experiment discussed in Problem 6, Carly's bag captured 0.0052 mol $CO_2$ in Trial B. Calculate the percent yield for this trial.

9. In the experiment discussed in Problem 6, Carly's bag captured 0.0035 mol $CO_2$ in Trial C. Calculate the percent yield for this trial.

# 11 What Is in My Blood?

## LEARNING GOALS

- **To gain experience with molar conversions**
- **To use molar and concentration conversions to interpret human physiological data**
- **To practice parts per million calculations**

## INTRODUCTION

Your blood is a complex mixture. Some of its components, such as salt ions, are very small. Some of its components, such as blood cells, are much larger. We are going to consider several of these components, paying particular attention to the amount and concentration of each in your blood.

## PROBLEMS

1. The average human contains 4.7 L of blood, which has a sodium concentration of 139 mM and a potassium concentration of 4.8 mM (1000 mM = 1 M).

   a) How many grams of sodium are contained in the average human's blood?

   b) How many grams of potassium are contained in the average human's blood?

2. Normal saline solution is used in intravenous drips for patients who cannot take fluids by mouth. This solution is made by dissolving 9.00 g sodium chloride (NaCl) in water and diluting it to 1000 mL.

   a) Calculate the sodium concentration in normal saline solution in millimolar (mM).

   b) Compare the concentration of sodium in human blood from problem 1 with that of normal saline solution.

3. One common intravenous therapy involves the administration of Hartmann's solution to replace lost body fluid and mineral salts. Hartmann's solution is a solution with several salts, including potassium chloride (KCl), calcium chloride ($CaCl_2$), sodium chloride (NaCl), and sodium lactate ($NaC_3H_5O_3$).

   a) Determine the mass in grams of each solute required to make 750 mL of Hartmann's solution with the following composition:

   5.00 mM   KCl

   2.00 mM   $CaCl_2$

   102 mM   NaCl

   29.0 mM   $NaC_3H_5O_3$

b) What is the final molar concentration of sodium ions and chloride in Hartmann's solution?

4. Many solutions in the laboratory are prepared at concentrations greater than what is needed for an application. These concentrated liquids are called stock solutions and are diluted with water to reach the desired final concentration. Determine how many mL water and stock solution are needed to prepare 750 mL of Hartmann's solution. The stock solution contains four salts as shown below.

Stock solution:   25.0 mM    KCl

                          10.0 mM    $CaCl_2$

                          510 mM     NaCl

                          145 mM     $NaC_3H_5O_3$

5. The average adult contains 5.1 erythrocytes (red blood cells) per microliter of blood. What is the molar concentration of erythrocytes in human blood?

6. The average diameter of an erythrocyte is 7.7 $\mu$m. How many meters would the erythrocytes from the average human span if placed end to end? (For reference, an American-football field is approximately 91 m in length!)

7. The average density of whole human blood is 1.06 g/mL and the sodium and potassium concentrations were given in Problem 1.

   a) What is the concentration of sodium in the blood expressed in parts per million?

   b) What is the concentration of potassium in the blood expressed in parts per million?

## Reference

Blood volume data, Problem 1: Bauer, J. D.; Ackermann, P. G.; Toro, G. Eds. *Clinical Laboratory Methods,* 8th ed.; C.V. Mosby Company: St. Louis, MO, 1974.

Written by Theodore Gries.

# How Can a Standard Curve Be Used to Monitor Phosphate in the Environment?

## LEARNING GOALS

- To practice molarity calculations

- To practice dilution calculations

- To practice fitting trend lines to data for calibration curves

## INTRODUCTION

Phosphate is an important nutrient in aqueous ecosystems, but it can also be a pollutant. In fact, high concentrations of phosphate lead to eutrophication, which depletes lakes and streams of oxygen and leads to fish kill. Sources of anthropogenic phosphates include fertilizer, animal waste, and wastewater effluent. To test for phosphate, environmental scientists often compare samples against standards of known phosphate concentration. The phosphate standards are used to create a calibration curve from which the concentration of an unknown sample can be estimated. Interestingly, environmental chemists denote concentration in units of parts per billion (ppb) and prepare their calibration curves with respect to the phosphorus (P) concentration rather than the phosphate ($PO_4^{3-}$) concentration. In this activity, you will learn how to prepare standards of known concentration as an environmental chemist would and how to interpret data collected from environmental samples.

## PART I: SOLUTION AND DILUTION PROBLEMS

1. Imagine that you are an environmental chemist analyzing phosphate concentrations in a river sample. You need to prepare phosphate standards for a calibration curve using the glassware described here:

- Class A volumetric flasks: 50.00, 100.0, 250.0, and 1000 mL

- Class A volumetric pipettes: 1.00, 2.00, 5.00, 10.00, 25.00, and 50.00 mL

- Analytical balance valid to four places to the right of the decimal

a) Describe how you would prepare a 100 mg/L phosphorus stock solution using monopotassium phosphate $(KH_2PO_4)$. You can do this either by preparing a 100 mg/L solution directly, or by first preparing a more concentrated standard (e.g., 1000 mg/L) and diluting it. Realize that you do not want to use too much $KH_2PO_4$ (it would be wasteful to use more than 0.5 g $KH_2PO_4$). Also, if you use less than 0.01 g $KH_2PO_4$, you will have trouble getting an accurate measurement, so this is not recommended.

b) Describe how you would prepare a 200 $\mu$g/L (ppb) phosphorus standard from your 100 mg/L phosphorus stock solution.

c) Describe how you would prepare a 150 $\mu$g/L (ppb) phosphorus standard from your 200 $\mu$g/L phosphorus stock solution.

d) Describe how you would prepare a 100 $\mu$g/L (ppb) phosphorus standard solution.

e) Describe how you would prepare a 50.0 $\mu$g/L (ppb) phosphorus standard solution.

2. Now imagine that you have created a 102 mg/L phosphorus stock solution rather than the suggested 100 mg/L phosphorus stock. Being slightly off in your mass is not a problem and will often make labs much faster because you won't waste valuable time at the balance, but you will need to adjust your calculations accordingly. What will the concentration of the ~200 $\mu$g/L phosphorus standard solution be in this case? (We're calling it an ~200 $\mu$g/L phosphorus solution because the solution is close to but slightly off this concentration.)

## PART II: CALCULATION OF PHOSPHORUS CONCENTRATION IN ENVIRONMENTAL SAMPLES

3. Using the ascorbic acid method, you measure the absorbance of your standards using visible spectroscopy (see the following table). Use a spreadsheet program or a graphing calculator to fit a trend line to the data (x-axis: concentration; y-axis: absorbance).

| Phosphorus Concentration ($\mu$g/L) | Absorbance |
|---|---|
| 50.0 | 0.094 |
| 100 | 0.312 |
| 150 | 0.520 |
| 200 | 0.862 |
| River sample 1 | 0.345 |
| River sample 2 | 0.353 |
| River sample 3 | 0.348 |

Write the equation and the $r^2$ value of the trend line here. The square of the correlation, $r^2$, is a measure of how well the regression explains the data.

4. Use the equation of the line to calculate the phosphorus concentration in the river samples, and report the concentration values here.

5. What is the average phosphorus concentration for the river samples?

6. The U.S. Environmental Protection Agency (EPA) measured the phosphorus concentration of the river to be 110 $\mu$g/L. Calculate the percent error using the following equation:

$$\text{percent error} = \frac{(\text{experimental value} \, - \, \text{true value})}{\text{true value}} \times 100\%$$

7. Describe two sources of error that might account for the deviation of your phosphorus concentration from the EPA-measured value. Be sure to account for the correct sign of the percent error.

# Laboratory: How Can We Group Household Chemicals Using Probes?

- **To carefully observe and interpret laboratory experiments**
- **To recognize acids and bases**

## INTRODUCTION

To explain their observations, chemists often look for ways to group or classify compounds. One historical classification of substances is "sour" or "slimy." Another classification system uses chemical probes such as phenolphthalein, bromothymol blue, or cabbage-juice indicators. This experiment requires you to organize, systematize, and interpret a set of observations. Although the experiment appears to ask you simply to categorize common household products as sour or slimy, it also requires that you do so in an iterative way. Based on *past* experience, do any of the products you're testing taste sour? (*Caution:* Never taste anything in a laboratory.) How do the chemical probes respond to those items? Based on *past* experience, do any of the products on the next page feel slimy? (*Caution:* Never touch chemicals in a laboratory.) Do the chemical probes respond differently to these sour and slimy products?

## PROCEDURE

- Work in pairs.
- Record your observations using the table provided here or a similar table in your laboratory notebook.
- Make note of your past experiences (slimy, sour) with each chemical.
- Obtain a spot plate or place a sheet of wax paper over a spare copy of the data table. Add a drop of the respective household products in each of the appropriate data table cells. Record your initial observations of color in the table.
- Add a drop of chemical probe to the sample drop, being careful not to contaminate the indicator dropper by touching the sample. Record in the table any *changes* in the solution *other than those due to dilution*. For example, if the final solution is

**47**

yellow, is the color the same or lighter yellow as the original? If so, record "no change." (A lighter yellow color could simply be due to dilution). If the yellow is different, record, for example, "darker yellow."

■ Which household chemicals give the same responses to the probe solutions?

| Chemical | Past Experience: Sour or Slimy? | Initial Color Observation | Phenolphthalein Probe | Bromothymol Blue Probe | Cabbage-Juice Probe |
|---|---|---|---|---|---|
| Lemon juice | | | | | |
| Dill pickle juice | | | | | |
| Glass Cleaning Solution | | | | | |
| Liquid hand soap | | | | | |
| Vinegar | | | | | |
| Ammonia | | | | | |
| Baking soda (in water) | | | | | |
| Bathroom tile and tub cleaning solution | | | | | |
| Pineapple juice | | | | | |

## DATA ANALYSIS QUESTIONS

1. What color changes are observed for phenolphthalein?

2. What color changes are observed for bromothymol blue?

3. What color changes are observed for cabbage juice?

4. Which household chemicals give the same color responses to the probe solutions?

5. For each chemical probe, what color is it when exposed to a sour substance?

6. For each chemical probe, what color is it when exposed to a slimy substance?

7. Of the products that you're not familiar with, which products would you expect to taste sour? Which would you expect to feel slimy?

8. Do any of the indicators give conflicting results for a particular household product? What does that tell you about the characteristics of that product?

9. You have classified some chemicals as belonging to one of two different groups, sour or slimy. Instead of this original historical classification, the sour group is now called acids and the soapy/slimy group is called bases. (The original terminology lingers in the German language: The word for *acid* in German is *Säure*. An English word with the same root is *sauerkraut*.)

a) Which chemicals would you classify as acids?

b) Which chemicals would you classify as bases?

---

Adapted from the ChemConnections module "Soil Equilibria: What Happens to Acid Rain?" by George Lisensky, Roxanne Hulet, Michael Beug, and Sharon Anthony.

---

# What Can Titration Tell Me about My Food?

### LEARNING GOALS

- **To practice titration calculations**
- **To learn to use statistics to analyze titration data**

## INTRODUCTION

Titration is a powerful quantitative technique often used by chemists to determine the concentration of a solute in solution. A *titrant*, a reagent of known concentration, is reacted with an *analyte* of unknown concentration. A colored product or chemical indicator can be used to determine the equivalence point at which the titration is complete, when the number of moles of the titrant is equal to the number of moles of the analyte, or some multiple thereof according to the balanced chemical equation. There are many types of titrations, and common titration techniques include acid–base titration and reduction–oxidation titration. Environmental chemists routinely use titration to determine the concentration of dissolved oxygen and of chloride ion in water samples. Titration can also give us information about the concentration of nutrients in our food. This activity will give you practice in analyzing titration data.

## PROBLEMS

1. Vitamin C is an essential nutrient found in citrus fruits and peppers, among other foods. The chemical structure of vitamin C ($C_6H_8O_6$) is

   HO     OH

   O       CHOHCH$_2$OH
        O

   A titration can be used to determine the concentration of vitamin C in foods. Iodine is often used as the titrant. The balanced equation for this reaction is

   $$C_6H_8O_6(aq) + I_2(aq) \rightarrow C_6H_6O_6(aq) + 2\,I^-(aq) + 2\,H^+(aq)$$

**51**

a) Which has more vitamin C, a red bell pepper or an orange? To determine this, one could use a 0.002 M iodine solution to titrate the juice extracted from each. The following table includes data from such an experiment. On a separate sheet of paper, use the data in this table to calculate whether a red bell pepper or an orange contains more vitamin C. Record your results in the table below.

**Titration Data for Oranges and Red Bell Peppers**

|  | Orange | Red Bell Pepper |
|---|---|---|
| Volume of juice extracted (mL) | 175 | 130 |
| Amount of juice titrated (mL) | 5.0 | 5.0 |
| Initial volume of $I_2$ (mL) | 0.0 | 6.2 |
| Final volume of $I_2$ (mL) | 6.2 | 14.8 |
| Volume of $I_2$ solution (mL) | | |
| Volume of $I_2$ solution (L) | | |
| Amount of $I_2$ added (mol) | | |
| Amount of vitamin C (mol) | | |
| Mass of vitamin C (mg) | | |
| Concentration of vitamin C in juice (mg/mL) | | |
| Mass of vitamin C in each fruit (mg) | | |

b) According to your calculations, which has more vitamin C, orange juice or red pepper juice? Does this surprise you?

2. Historically, vinegar played an important role in the preservation of food because of the acidity and antimicrobial properties of its main component, acetic acid. There are many types of vinegar found at your local grocery store, including balsamic, rice, red wine, and apple cider vinegars. A titration will help you determine the concentration of acetic acid ($CH_3COOH$) in vinegar.

The balanced equation for the titration of acetic acid in vinegar with sodium hydroxide (NaOH) is

$$CH_3COOH(aq) + NaOH(aq) \rightarrow H_2O(l) + CH_3COONa(aq)$$

Assume you titrated a sample of vinegar using 0.145 M sodium hydroxide. Use the data in the following table to determine the vinegar's acetic acid concentration in moles per liter.

| Titration Data for Vinegar | |
|---|---|
| | TITRATION DATA |
| Volume of vinegar titrated (mL) | 5.00 |
| Initial volume of NaOH (mL) | 0.00 |
| Final volume of NaOH (mL) | 29.32 |

Most vinegars are 4% to 6% acetic acid. Are your results within this range?

3. Oxalic acid ($H_2C_2O_4$), found in dark, leafy, green vegetables such as spinach, can inhibit iron absorption in the body. To determine the concentration of oxalic acid in a food, one can titrate a sample with a permanganate solution, which changes color on reaching the equivalence point. If one cup of spinach contains 180 mg of oxalic acid, how many milliliters of 0.0100 M permanganate solution would be required to reach the equivalence point? The balanced equation is given here.

$$H_2C_2O_4(aq) + MnO_4^-(aq) \xrightarrow{H^+} CO_2(g) + Mn^{2+}(aq)$$

4. Stan performs three vitamin C titrations of orange juice. He finds vitamin C concentrations of 0.35 mg/mL, 0.32 mg/mL, and 0.36 mg/mL.

a) Find the mean of these concentrations.

b) Find the standard deviation of these concentrations.

c) Find the (dimensionless) coefficient of variation (CV) for these three values.

$$\text{coefficient of variation} = \frac{\text{standard deviation}}{\text{mean}} \times 100\%$$

d) If the coefficient of variation is greater than 5%, Stan's professor insists that he perform more titrations so he can trust his titration data. Is he finished, or does he need to titrate again?

# What Reduction–Oxidation Reactions Happen in Nature?

**LEARNING GOALS**

- To identify reduction and oxidation in chemical reactions

- To assign oxidation numbers to atoms in reduction and oxidation reactions

## INTRODUCTION

Reduction and oxidation drive many of the reactions we notice every day, including rusting, photosynthesis, and respiration. This activity highlights some of these reduction–oxidation reactions and asks you to assign oxidation numbers to atoms in various reduction–oxidation reactions.

## PROBLEMS

1. The following reaction happens during the smelting of copper ore. Label the oxidation number for each of the atoms, and describe which species are oxidized and which are reduced.

$$Cu_2S(s) + O_2(g) \rightarrow 2\,Cu(s) + SO_2(g)$$

2. The following reaction happens during the rusting process. Label the oxidation number for each of the atoms, and describe which species are oxidized and which are reduced.

$$4\,Fe(s) + 3\,O_2(g) \rightarrow 2\,Fe_2O_3(s)$$

3. Groundwater has low levels of dissolved oxygen because it has not been exposed to air for long periods. It also tends to have high levels of soluble $Fe^{2+}$. When groundwater reaches the surface and oxygen dissolves in it, soluble $Fe^{2+}$ is oxidized to insoluble $Fe^{3+}$, forming a deposit of iron(III) hydroxide, $Fe(OH)_3$. Label the oxidation number for each of the atoms involved in this reaction, and describe which species are oxidized and which are reduced.

$$4\ Fe^{2+}(aq) + O_2(g) + 2\ H_2O(l) + 8\ OH^-(aq) \rightarrow 4\ Fe(OH)_3(s)$$

4. Cellular respiration occurs via the reaction given here. Label the oxidation number for each of the atoms, and describe which species are oxidized and which are reduced.

$$C_6H_{12}O_6(aq) + 6\ O_2(g) \rightarrow 6\ CO_2(g) + 6\ H_2O(l)$$

5. Agricultural areas tend to have high concentrations of nitrate ($NO_3^-$) in the groundwater and surface water due to runoff of fertilizers. Blue baby syndrome, also called *methemoglobinemia*, is a problem for babies who drink water with too much nitrate. The reaction for the process that causes this syndrome is given here. Bacteria reduce nitrate to nitrite ($NO_2^-$), which combines with hemoglobin in the blood, preventing hemoglobin from absorbing oxygen and transferring oxygen to cells. Label the oxidation number for each of the atoms, and describe which species is reduced.

$$NO_3^-(aq) + 2\ H^+(aq) + 2\ e^- \rightarrow NO_2^-(aq) + H_2O(l)$$

6. The decay of organic material in anaerobic environments accounts for most global methane production. During a disproportionation reaction, a carbohydrate (e.g., glucose, $C_6H_{12}O_6$) is simultaneously oxidized and reduced to form carbon dioxide and methane ($CH_4$). This process occurs in wetlands, rice paddies, domestic sewage, and landfills, among other places. (A similar process called *enteric fermentation* occurs in the digestive tracts of termites and ruminants, such as cattle.) Label the oxidation number for each of the atoms, and describe which species are oxidized and which are reduced.

$$C_6H_{12}O_6(aq) \rightarrow 3\ CO_2(g) + 3\ CH_4(g)$$

7. Nitrous oxide ($N_2O$), a greenhouse gas, is produced by anaerobic bacteria in the presence of water by the following reaction. Label the oxidation number for each of the atoms, and describe which species are oxidized and which are reduced.

$$2\,NO_3^-(aq) + H_2O(l) \rightarrow N_2O(g) + 2\,O_2(g) + 2\,OH^-(aq)$$

8. Nitrous oxide is removed from the atmosphere via a reaction with high-energy oxygen atoms as shown here. Label the oxidation number for each of the atoms, and describe which species are oxidized and which are reduced.

$$N_2O(g) + O(g) \rightarrow 2\,NO(g)$$

9. Catalytic converters remove carbon monoxide and nitric oxide, NO, from automobile exhaust as shown here. Label the oxidation number for each of the atoms, and describe which species are oxidized and which are reduced.

$$2\,NO(g) \rightarrow N_2(g) + O_2(g) \qquad\qquad 2\,CO(g) + O_2(g) \rightarrow 2\,CO_2(g)$$

# How Is Copper Processed?

- **To practice assigning oxidation numbers**
- **To practice balancing half-reactions**

## INTRODUCTION

With a history of use stretching back over 10,000 years, copper is one of the most widely used metals in the world. Copper's high thermal and electrical conductivity, malleability, and beautiful luster, has led to its use in everything from cooking pots to power lines. Copper is present in the Earth's crust primarily as sulfide and oxide minerals at concentrations of less than 1%. This activity will help you discover how copper is extracted from these minerals and converted to its metallic form.

## PART I: ASSIGNING OXIDATION NUMBERS

When mined, copper ore contains a wide range of contaminant metals that must be taken into consideration when copper is extracted from the ore. Copper, like many other metals, is found with multiple oxidation states.

1. Write the oxidation number for each of the elements in the following table of minerals related to copper mining.

| Mineral | Formula | OXIDATION STATE | | | | | |
|---------|---------|------|------|------|------|------|------|
| | | Cu | Fe | O | S | Al | Si |
| Troilite | $FeS$ | — | | — | | — | — |
| Bornite | $Cu_5FeS_4$ | | | | | | |
| Malachite | $Cu_2CO_3 \cdot Cu(OH)_2$ | | — | | — | | — |
| Halotrichite | $FeAl_2(SO_4)_4 \cdot 22(H_2O)$ | — | | | | | — |
| Iron Pyrite | $FeS_2$ | — | | — | | — | — |
| Brochantite | $CuSO_4 \cdot 3Cu(OH)_2$ | | — | | | — | — |
| Jarosite | $KFe_3(SO_4)_2(OH)_6$ | — | | | | | — |
| Tenorite | $CuO$ | | — | | — | — | — |
| Chrysocolla | $CuSiO_3 \cdot 2H_2O$ | | — | | — | — | |

Copper-containing ore must be processed to extract metallic copper. This is a multistep process that often entails heating the ore to high temperatures during the smelting process, exposing the ore to oxygen-rich air to oxidize impurities such as iron, and using electrochemical methods to recover the dissolved copper.

2. Matte smelting occurs when the ore is exposed to an oxidizing atmosphere at temperatures near 1200°C. The smelting reaction for chalcopyrite ($CuFeS_2$) is given here:

$$2\,CuFeS_2(s) + 5/2\,O_2(g) \rightarrow Cu_2S(s) + FeO(s) + FeS(s) + 2\,SO_2(g)$$

a) Label the oxidation number for each species involved in the reaction.

b) Which species were oxidized?

c) Which species were reduced?

3. Another step in the processing of copper-containing ore is shown here for chalococite ($Cu_2S$).

$$Cu_2S(s) + O_2(g) \rightarrow 2\,Cu(s) + SO_2(g)$$

a) Label the oxidation state for each species involved in the reaction.

b) Which species were oxidized?

c) Which species were reduced?

## PART II: BALANCING OXIDATION/REDUCTION REACTIONS

Balance the pertinent redox reactions here.

4. One of the hazards of copper mining is acid mine drainage. When sulfide-based ores are exposed to oxygen and water, acid is produced.

$$\_FeS_2(s) + \_O_2(g) + \_H_2O(l) \rightarrow \_Fe(OH)_3(s) + \_SO_4^{2-}(aq) + \_H^+(aq)$$

5. Acid mine drainage is caused predominantly by the physical and chemical weathering of iron pyrite (iron(II) sulfide) in the presence of oxygen and water. You may know this mineral as "fool's gold."

$$\_FeS_2(s) + \_O_2(g) + \_H_2O(l) \rightarrow \_Fe^{2+}(aq) + \_SO_4^{2-}(aq) + \_H^+(aq)$$

6. Iron(II) ions, formed in the weathering of iron pyrite, are oxidized to form iron(III) ions, as shown in the following reaction.

$$\_Fe^{2+}(aq) + \_O_2(g) + \_H^+(aq) \rightarrow \_Fe^{3+}(aq) + \_H_2O(l)$$

## PART III: ADDITIONAL OXIDATION/REDUCTION REACTIONS

Assume acidic conditions and balance the following reactions, which are also relevant to copper mining.

7. $CuS(s) + Fe^{3+}(aq) \rightarrow Cu^{2+}(aq) + Fe^{2+}(aq) + SO_4^{2-}(aq)$

8. $Fe^{2+}(aq) + O_2(g) \rightarrow Fe^{3+}(aq)$

9. $FeS(s) + ClO_3^-(aq) \rightarrow Fe^{2+}(aq) + Cl^-(aq) + SO_4^{2-}(aq)$

10. $Cu_2S(s) + ClO_3^-(aq) \rightarrow Cu^{2+}(aq) + SO_4^{2-}(aq) + Cl^-(aq)$

Adapted by Kevin Braun, Sharon Anthony, and Mary Walczak, from the ChemConnections module "Should We Recommend Building a Copper Mine?" by Mary Walczak, Linda Zarzana, Doug Williams, and Paul Charlesworth.

# How Much Heat Is Released through Fuel Combustion?

- **To introduce calorimetry as a means of measuring the energy content of fuels**

- **To determine the approximate energy per mole released for common fuels**

- **To perform mathematical calculations related to energy and percent error**

## INTRODUCTION

A traditional automobile engine propels a car by converting heat produced through fuel combustion into work. The efficiency of this conversion is dependent on many elements of the engine design and the choice of fuel. A good fuel must release energy when it combusts—and the more energy, the better!

In this activity, you will set about the task of determining quantitatively the enthalpy change of combustion, $\Delta H_{combustion}$, of some common components of gasoline. The compounds you will explore include isooctane—$(CH_3)_3CCH_2CH(CH_3)_2$, a common gasoline component—and two fuel oxygenators, ethanol—$C_2H_5OH$ and methyl tert-butyl ether (MTBE)–$CH_3OC(CH_3)_3$. Oxygenators are oxygen-containing molecules that are added to gasoline to facilitate a more complete combustion reaction, thus reducing the production of carbon monoxide. In the United States, MTBE has been phased out (because of health risks associated with its ready dissolution in potable water) and replaced with ethanol. However, MTBE is still used extensively around the world because of its cheap production cost. Your goal is to determine whether there is a thermodynamic cost associated with the switch from MTBE to ethanol.

The heat released during combustion can be measured experimentally using a method called **calorimetry**. In this method, a container of water is heated by combusting a fuel in a burner, and the temperature change of the water and the mass of fuel combusted are recorded. According to the first law of thermodynamics, the energy gained by the container and the water in it should equal the energy given off by the combustion reaction, assuming you can correct for any other heat loss to the surrounding environment. The energy absorbed by the container and water is calculated by using the temperature change and the

**63**

known **heat capacity** of the container and water. When the container of water is open to the atmosphere, thus working under constant pressure, the heat absorbed by the container and water can be equated to the enthalpy change ($\Delta H$) of the system, which is measured in kilojoules per mole (kJ/mol). Suppose you were to construct a fuel calorimeter by suspending a soda can containing approximately 200 mL of water from a ring stand over a fuel burner. You place a thermometer in the can and surround the apparatus with an aluminum baffle to prevent drafts.

Your first step is to calibrate the calorimeter using methanol ($CH_3OH$), a fuel with a known enthalpy of combustion. The heat loss factor ($f$) of the calorimeter accounts for the fact that not all the heat generated by the combusted fuel goes into the calorimeter, which consists of the can and water where the temperature change is measured. Where else can the heat go?

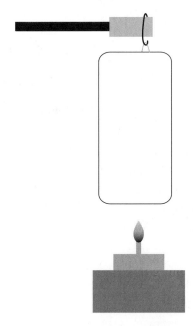

Calorimeter setup for determining heats of combustion (aluminum baffle and thermometer not shown).

The actual heat generated by the fuel is calculated from the moles of fuel consumed, thus the fuel burner is weighed before and after each experiment. For reproducibility, the methanol is combusted long enough to produce at least a 20°C rise in the water. Data and observations from three methanol-combustion trials are summarized in the first table in the following Data section.

Suppose you next use the calorimeter and its calibration data to determine the heat of combustion for the three unknown fuels. Data from these experiments are summarized in the second table of the Data section. In this second experiment, suppose you notice some stark differences between the characteristics of the flames associated with the combustion of the three fuels and take note of them in your data table. Why is it important to take such detailed observations when you conduct an experiment? Using the data in the tables here, calculate the calorimeter's heat loss factor and the heats of combustion for the three fuels.

## DATA

### Data for Determination of the Calorimeter's Heat Loss Factor, f

| Trial | Mass of H$_2$O (g) | Burner Mass Initial (g) | Burner Mass Final (g) | $T_{Initial}$ (°C) | $T_{Final}$ (°C) | Observations |
|---|---|---|---|---|---|---|
| 1 | 198.10 | 145.82 | 144.30 | 21.0 | 42.2 | Colorless flame, small amount of soot. |
| 2 | 198.10 | 144.30 | 142.76 | 35.6 | 56.2 | Colorless flame, small amount of soot. |
| 3 | 198.10 | 142.76 | 141.24 | 48.2 | 70.3 | Small amount of soot observed on the bottom of the can. |

### Data for Determination of the Fuel Enthalpy of Combustion of Ethanol, Isooctane, and MTBE. Values Are the Average of Three Trials for Each Fuel

| Fuel | Mass of H$_2$O (g) | Burner Mass Initial (g) | Burner Mass Final (g) | $\Delta T$ (°C) | Observations |
|---|---|---|---|---|---|
| Ethanol | 196.60 | 155.66 | 154.60 | 20.4 | Light blue flame, more soot than methanol. |
| Isooctane | 199.40 | 138.49 | 137.68 | 21.8 | Thick, dark soot produced. Can covered in soot. |
| MTBE | 197.10 | 140.96 | 139.97 | 22.0 | Can replaced. Thick, dark soot produced. Can again covered in soot. |

## CALCULATIONS

### Calculating the Heat Loss Factor, f

The heat loss factor is the absolute value of the ratio of the heat generated by burning $n$ moles of fuel, $Q_f$, to the heat absorbed by the calorimeter, $Q_c$, as shown by the following equation:

$$f = \left| \frac{\text{heat generated from fuel, } Q_f}{\text{heat absorbed by calorimeter, } Q_c} \right|$$

Ideally, the calorimeter would absorb all the heat generated, so $f = 1$. However, in reality some heat is lost to the surroundings, so the denominator is smaller than the numerator and $f > 1$, typically between 2 and 4. Answer the following questions to determine the heat loss factor for your calorimeter.

1. Use the data in the Data section to complete the following table, which will be used to calculate the heat generated by combustion of the methanol and the heat absorbed by the calorimeter.

| Trial | Mass of Methanol Burned (g) | Moles of Methanol Burned | $\Delta T$ (°C) |
|---|---|---|---|
| 1 | | | |
| 2 | | | |
| 3 | | | |

2. The heat generated by burning $n$ moles of methanol, $Q_f$, is calculated by multiplying the moles of methanol combusted by methanol's enthalpy of combustion ($\Delta H_{comb} = -726$ kJ/mol).

$$Q_f = \Delta H_{comb} \times n_{methanol}$$

Calculate $Q_f$ for the three methanol-combustion trials.

| Trial | $Q_f$ [kJ] |
|-------|-----------|
| 1 | |
| 2 | |
| 3 | |

3. Assuming the heat absorbed by the can is negligible, the heat absorbed by the calorimeter is calculated from the specific heat of water ($c_{H_2O} = 4.184$ J/(g•K)) multiplied by the mass of the water ($m_{H_2O}$) multiplied by the temperature change of the calorimeter ($\Delta T_{calorimeter}$).

$$Q_c = (m_{H_2O} \times c_{H_2O}) \times (\Delta T_{calorimeter})$$

Calculate $Q_c$ for the three trials.

| Trial | $Q_c$ [kJ] |
|-------|-----------|
| 1 | |
| 2 | |
| 3 | |

4. Calculate the heat loss factor for the three trials (be sure your units are consistent for $Q_c$ and $Q_f$). What is the average heat loss factor for the calorimeter?

| Trial | f |
|-------|---|
| 1 | |
| 2 | |
| 3 | |
| Average | |

## Calculating the Fuels' Enthalpy of Combustion

To calculate the molar enthalpy of combustion for the fuels, first calculate the heat absorbed by the calorimeter using the equation given in problem 3 and the fuel burn data in the second data table. Multiply $Q_c$ by the heat loss factor, calculated previously, and divide by the moles of fuel combusted, $n$.

$$\Delta H_{combustion} = \frac{f \times Q_c}{n}$$

5. Calculate the molar enthalpy of combustion in kilojoules per mole for the three fuels (note that the enthalpy change is the negative of the heat of combustion as defined previously).

| Fuel | $Q_c$ [kJ] | $n$ [mol] | $\Delta H_{combustion}$ [kJ/mol] |
|------|-----------|-----------|----------------------------------|
| Ethanol | | | |
| Isooctane | | | |
| MTBE | | | |

## ANALYSIS

6. Based on your results for the molar enthalpies of combustion of ethanol, isooctane, and MTBE, would you expect the enthalpy of combustion of methane ($CH_4$), a major component of natural gas, to be greater than or less than that of ethanol? Explain.

7. How do your experimentally measured enthalpy of combustion values compare with the literature values reported here? Calculate a percent error for each fuel.

$$\text{percent error} = \frac{|(\text{experimental value} - \text{true value})|}{\text{true value}} \times 100\%$$

| Fuel | Literature Value (kJ/mol) | Percent Error |
|------|---------------------------|---------------|
| Ethanol | −1366.8 | |
| Isooctane | −5455.5 | |
| MTBE | −3359.7 | |

8. List three possible sources of error in this experiment that would account for any deviation from the literature value. (*Hint:* What do the observations tell us about the combustion reaction?) How could you redesign this experiment to minimize these sources of error?

9. With the phase-out of MTBE, is there a thermodynamic cost to switching to ethanol? How will this affect the fuel efficiency (miles per gallon) of automobiles?

## Reference

Rettich, T.; Battino, R.; Karl, D. J. Heating Value of Fuels, *Journal of Chemical Education* **1988,** *65,* pp 554–555.

Adapted by Kevin Braun and Brock Spencer from the ChemConnections module "Can We Reduce Air Pollution from Automobiles?" by Howard Drossman and Wayne Tikkanen.

# Laboratory: How Much Energy Is in My Food?

## LEARNING GOALS

- **To introduce calorimetry as a means of measuring the energy content of food**

- **To design a laboratory experiment capable of determining the approximate energy per unit mass released by the combustion of a food substance**

- **To perform mathematical calculations related to energy and percent error**

## INTRODUCTION

Most of the energy we get from our food is supplied by carbohydrates and fats. These food molecules are broken down in our cells in oxidation reactions. In the presence of excess oxygen, the food molecules react in a complex series of reactions to produce carbon dioxide ($CO_2$) and water ($H_2O$). We breathe in oxygen to sustain this process and exhale carbon dioxide and water vapor as reaction products. The energy released in these reactions fuels our basal metabolism (the chemical reactions constantly occurring to maintain basic body functions) and our physical activity. Because the products of these oxidation reactions are the same when food is burned in a calorimeter as when it is metabolized in the body, we can approximate the energy value of carbohydrates and fats in our bodies by doing a simple calorimetry experiment.

Food chemists use a bomb calorimeter to measure the energy evolved from the combustion of food. A bomb calorimeter is a sealed steel vessel, or "bomb," designed to withstand high pressures and surrounded by a known volume of water—a sealed container at constant volume. The quantity of energy (measured in joules or calories) produced by the combustion of a known amount of food inside the bomb can be calculated from the temperature increase of the water. It should be noted that this approximation does not work as well for proteins because their nitrogen is released as $N_2$ in a bomb calorimeter, whereas the body produces mainly urea ($(NH_2)_2CO$). In addition, laboratory measurements of the energy content of foods tend to be higher than the energy obtained by the body because a small percentage of food is not digested or absorbed as it moves through the digestive tract. This percentage varies depending on the efficiency of a person's digestive system. Finally, a portion of the food's energy must be used for the process of digestion and metabolism.

## PURPOSE

The purpose of this investigation is to quantitatively determine the energy per unit mass released by the combustion of nuts, cheese puffs, or other snack foods.

## METHOD

A minimal amount of materials will be provided. You are responsible for designing your apparatus and setting the parameters for the experiment.

## PROCEDURE

Work in pairs. Using a paper clip and a cork (or rubber stopper), construct a "nut support stand" to hold your burnt offering. Fill a soda can with 200 mL of water. Suspend the soda can above the nut support stand. (Use your imagination!) Carefully set the food on fire and measure the temperature change in the water that results from the heat given off. Stir the water thoroughly to be sure that it has a uniform temperature. *Do not* leave the thermometer in the can while the nut is burning unless you can suspend the thermometer so it does not touch the bottom of the can. Try this experiment for at least one type of nut and a cheese puff. *Caution:* The ash left after burning the food is very sooty, so try not to get it on your clothes.

After conducting the experiment once, speculate on the sources of error. Will each of these errors increase or decrease the measured temperature change? If additional time is available, refine your apparatus and procedure to try to eliminate or account for this error. Because you are not burning the food in a sealed and insulated bomb calorimeter, your results will not be entirely precise. However, there are many ways to refine this experiment to decrease error. When you believe that you have the best apparatus and procedure possible, perform three or four additional trials to check the precision of your measurements.

In your laboratory notebook, make a data table that reports for each trial:

Exact volume of water

Initial temperature of water

Final temperature of water

Initial mass of food

Final mass of food

## USEFUL INFORMATION

One calorie is the amount of energy required to raise the temperature of 1.00 g of water 1.00°C. The density of water is 1.00 g/mL.

A chemist's calorie (lowercase *c*) is different from a food Calorie (capital *C*). A food Calorie is actually a chemist's kilocalorie.

$$1 \text{ Cal} = 1 \text{ kcal} = 1000 \text{ cal}$$

## QUESTIONS

1. Calculate the energy released per unit mass, in Calories per gram, for each food.

2. Is the energy liberated in the combustion of your food the entire amount of energy contained in the sample? Explain.

3. How does the energy released in your experiment compare with the value reported on the nutritional label or in a table of nutritional energy values? Calculate a percent error:

$$\text{percent error} = \left( \frac{\text{experimental value} - \text{value on label}}{\text{value on label}} \right) \times 100\%$$

4. List four possible sources of error in this experiment. Which sources are inherent in the technique? Speculate on a technique that would minimize these sources of error.

---

Adapted from the ChemConnections module "Would You Like Fries with That?: The Fuss about Fats in Our Diet" by Sandra Laursen and Heather Mernitz.

---

# How Can I Calculate the Energy Content of Fuels?

## LEARNING GOALS

- **To perform bond enthalpy calculations for various fuels**

- **To compare and rank fuels on the basis of their energy density**

## INTRODUCTION

In this activity, you will calculate the heat of several reactions using bond energies. In a hypothetical process, all the bonds in the reactant molecules are broken simultaneously and then reformed to produce the product molecules. The enthalpy change for the process of breaking all the bonds in the reactants is the sum of all the bond energies in the reactant molecules. The enthalpy change for forming the new bonds is the sum of all the bond energies in the product molecules. Remember that energy is needed to break bonds and energy is given off when bonds are formed. For a reaction that is exothermic (gives off energy), the total enthalpy change $\Delta H$ of the reaction is a negative number. A reaction with a positive $\Delta H$ is an endothermic reaction. Would a compound that undergoes an endothermic reaction make a good fuel? Why or why not?

## BOND ENTHALPY CALCULATIONS

Calculate $\Delta H_{combustion}$ for the fuels below based on bond energies. Remember to first balance the equation for the reaction of your fuel with oxygen to produce water and carbon dioxide.

Use a table of bond energies from your textbook or provided by your instructor to complete the following worksheet. Most tables of bond energies report values in kilojoules per mole (kJ/mol) because the joule is the SI unit of energy. You will also be asked to convert from kilojoules per mole to Calories per gram (with *Calories* written with a capital *C*). The unit of food Calories (1 Cal = 1 kcal = $10^3$ cal = 4.184 kJ) is often a more recognizable unit of energy in the United States, especially when it is used in relation to energy from food sources.

| Glucose | Octane | Stearic Acid | Ethanol |
|---------|--------|--------------|---------|

**Fuel:** Ethanol

**Balanced combustion equation (line-bond notation or Lewis structure):**

**Bonds broken:**

| Bond type (e.g. C–H) | Moles of bonds | Energy per mole of bond | Total energy |
|----------------------|----------------|-------------------------|--------------|

Enthalpy change due to bond breaking $= \Delta H$ (breaking bonds) $=$

**Bonds formed:**

| Bond type (e.g. C–H) | Moles of bonds | Energy per mole of bond | Total energy |
|----------------------|----------------|-------------------------|--------------|

Enthalpy change due to bond formation $= \Delta H$ (forming bonds) $=$

**Total enthalpy change for this fuel source:**

**Molar mass**

**Cal/g**

Using the format described previously, calculate the enthalpy change for the combustion of the three other fuel sources. Calculate your answers in both kilojoules per mole and Calories per gram.

    a) Stearic acid

b) Glucose

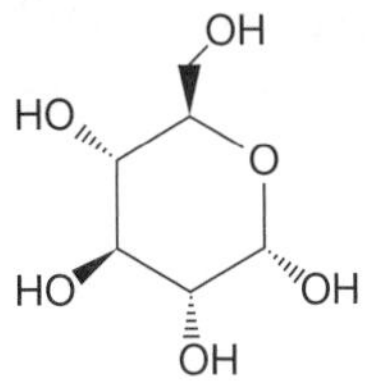

c) Octane

# QUESTIONS

Compile your results in the summary table below and use them to assist you in answering the following questions.

| Summary Table | | |
| --- | --- | --- |
| Fuel | kJ/mol | Cal/g |
| Ethanol | | |
| Stearic acid | | |
| Glucose | | |
| Octane | | |

1. The molecules above all share one thing in common. They are all used as fuels. Describe where you would find each of the chemicals above being used as a fuel.

2. Rank the fuels from highest to lowest energy output per mole ($\Delta H$) and energy output per gram on combusting.

3. Glucose is metabolized (combusted) by the body to provide quick energy. Considering bond energies, suggest a reason why oxygen is used in the metabolism of glucose, whereas nitrogen is inert.

4. Why must the stoichiometry of a reaction be known to estimate the enthalpy change of a reaction from bond energies?

5. Glucose is a simple carbohydrate, whereas stearic acid is a common fatty acid component of saturated fats. Protein, when metabolized, produces about the same amount of energy per gram as carbohydrates. Peanuts are typically 26% protein, 39% fat, and 22% carbohydrate by weight.

   a) Estimate the amount of energy produced in Calories when a 50 g bag of peanuts is metabolized.

   b) Calculate the number of Calories/gram of peanuts.

   c) If a typical diet contains 2000 Cal/day, how many pounds of peanuts would that be?

6. The energy obtained by metabolizing ethanol (alcohol) is sometimes referred to as "empty calories." Why might that be? How does its energy content per gram compare with that of proteins, carbohydrates, and fats?

Adapted by Heather Mernitz, Kevin Braun, Brock Spencer, and Kim Schatz from the ChemConnections module "Would You Like Fries with That? The Fuss about Fats in Our Diet" by Sandra Laursen and Heather Mernitz.

# How Does an Automobile Engine Convert Heat to Work?

- **To understand how an automobile engine works**

- **To understand how to calculate work from a pressure–volume diagram**

- **To introduce the Carnot cycle and explain why it provides the highest theoretical efficiency**

## INTRODUCTION

Unlike enthalpy, heat and work are not state functions. Variable amounts of heat or work can be obtained from any given chemical reaction, depending on the particular path or process from the initial to the final state. The first law of thermodynamics states that the total of the heat and the work, the total energy change, will be constant for a given change of state, but the amounts of heat and work can vary within that constant total energy change.

An interesting problem for automotive engineers is to optimize the conversion of the energy produced from fuel combustion into work used for moving an automobile. Though a "perfect" automotive engine would convert all of the energy from the fuel to work, such a completely efficient conversion is impossible according to the second law of thermodynamics.

A common type of work performed by chemical systems is expansion work, or pressure–volume (PV) work, due to gases in the chemical system pushing on the surroundings as they expand. For example, if you heat the gas inside a balloon, the gas expands, pushing back the atmosphere around the balloon and, in the process, doing work on the atmosphere. Note that the work done is not necessarily useful! However, the expansion could be made useful—the expanding balloon might lift a pencil sitting on top of it, for instance, or push a toy car sitting next to it. In a traditional car engine, energy from combustion heats the gas in the cylinder, raising its pressure and enabling it to expand and push a piston that is connected to the car's drive train.

As you work through this activity, you will see that the amount of work that can be done by a piston depends on how much the gas inside is allowed to expand while pushing the piston. Automotive engineers call this the compression ratio, defined as the ratio of the

**77**

maximum to the minimum volume defined by the piston's travel in an engine cylinder. Thus, compression ratio is directly proportional to the length of the piston stroke. A longer stroke gives a larger compression ratio, and a shorter stroke gives a smaller compression ratio. In this activity, you will consider how the compression ratio relates to the amount of work that can be done by the piston.

## FOUR-STROKE INTERNAL COMBUSTION AUTOMOBILE ENGINES

The vast majority of modern automobiles employ what is known as a *spark ignition piston engine* for propulsion. "Spark ignition" refers to the fact that the fuel is ignited by a spark from a spark plug. Energy is derived from the combustion in a four-step cycle, as illustrated in the figure here, but beware, the steps are not illustrated in sequential order.

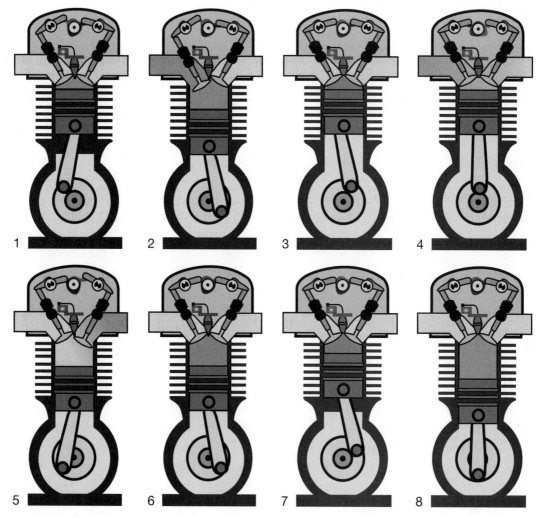

Schematic of a four-stroke spark ignition piston engine.

From the "Start your engines!" webpage developed by Matthias Geissbühler and Christoph Eggimann for ThinkQuest http://library .thinkquest.org/C006011/ Reprinted with permission of the authors.

A color version of this figure can be found at the website noted above.

1. The idealized model of a four-stroke engine is called the Otto cycle, named after the German engineer Nikolaus Otto, who built the first prototype in 1876. This model is illustrated in the preceding figure. Your instructor should post or distribute a copy of this figure in color for the class.

Match each picture in the figure to the step described in the following list. The green and dark grey colors represent fuel before ignition and the red, orange, and light grey colors represent ignited fuel. More than one picture may illustrate each step.

Step 1a (intake): The piston chamber takes in the fuel–air mixture through the open valve.

Step 2a (downstroke): The fuel evaporates and the air and fuel mix during the downstroke of the piston.

Step 2b (compression): The piston compresses the mixture on the upstroke.

Step 3a (ignition): As the piston reaches the top of its stroke, a spark from the spark plug ignites the fuel mixture.

Step 3b (expansion): The heated gas expands and does work.

Step 4 (exhaust): At the bottom of the power stroke, the piston returns to the top of the cylinder to expel exhaust gases.

Step 1b (refill): The piston chamber refills with a fresh fuel–air mixture.

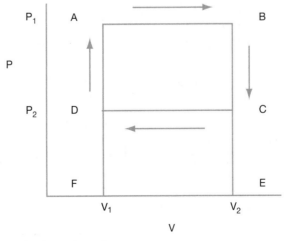

Pressure-Volume diagram for a simple stroke cycle.
From the "Start your engines!" webpage developed by Matthias Geissbühler and Christoph Eggimann for ThinkQuest http://library .thinkquest.org/C006011/ Reprinted with permission of the authors.

2. The amount of energy extracted from a gas can be calculated from the area under a PV curve. For the process A → B → C → D → A illustrated in the preceding figure, what is the total change of internal energy, assuming an ideal gas? (*Hint: Internal energy is a state function.*)

3. Given that work can be calculated as $P \times \Delta V$, why is no work done in process D → A and B → C in the pressure-volume diagram above.

4. Write expressions for $\Delta T$ in terms of $P$, $V$, and $R$ (universal gas constant) for 1 mol, $n$, of an ideal gas for the processes A → B and C → D. Explain whether heat is gained or lost in each step.

The theoretically most-efficient engine possible was described in 1824 by Sadi Carnot, a French engineer. He postulated, in his *Reflections on the Motive Power of Fire*, that no engine could convert 100% of heat to work and that the theoretically most-efficient heat engine could be described by an engine cycle illustrated in the figure below, which has two isothermal and two adiabatic steps. This ideal engine cycle is named the *Carnot cycle* in his honor.

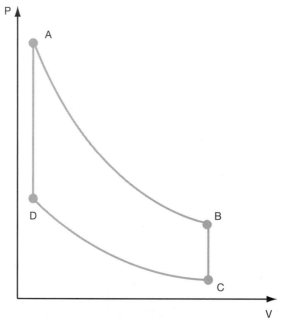

Carnot cycle for an ideal engine.

5. If a system goes through the cycle in order from A to D and returns to point A, what is the total change in internal energy? Given this, what is the relationship between the total amount of work and the heat added to the system?

6. Because the internal energy of a system is a function only of its temperature, what is the change in internal energy for an isothermal process?

7. According to the information in the Carnot cycle figure above, which two processes (A → B; B → C; C → D; D → A) are isothermal? Given your answer to question 4, what is the relationship between heat and work for these isothermal processes?

8. The efficiency of an engine can be calculated as the ratio of the work done to the heat added. Which of the four steps *adds* heat to the system?

9. The figure below represents the change in pressure and volume during the different steps in the four-stroke engine cycle. Match the letter of each PV diagram here to the appropriate picture in the first figure of the four-stroke internal combustion engine.

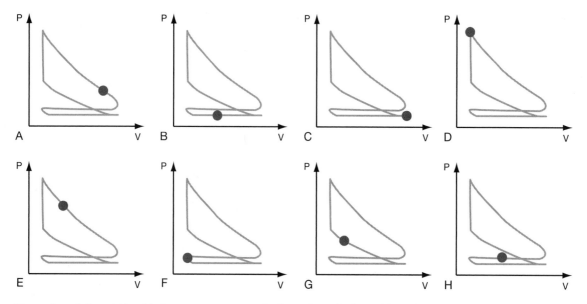

Illustration of the relationship between pressure and volume through the multiple cycles of a four-stroke engine.

From the "Start your engines!" webpage developed by Matthias Geissbühler and Christoph Eggimann for ThinkQuest http://library .thinkquest.org/C006011/ Reprinted with permission of the authors.

| PV Diagram | Engine Figure | PV Diagram | Engine Figure |
|---|---|---|---|
| A | | E | |
| B | | F | |
| C | | G | |
| D | | H | |

10. How would you calculate the work done by the engine illustrated in the preceding pressure–volume graphs?

11. How does the amount of work done by the engine represented in the pressure–volume graphs compare with the amount of work done by the Carnot engine?

Adapted by Howard Drossman from the ChemConnections module "Can We Reduce Air Pollution from Automobiles" by Sandra Laursen, Howard Drossman, and Wayne Tikkanen. Figures used with permission by Matthias Geissbuehler from animations available at http://library.thinkquest.org/C006011/.

# Laboratory: How Do Power Plants and Automobile Emissions Affect Lakes?

- **To practice experimental design**
- **To understand the chemistry of acid rain**

## INTRODUCTION

Acid rain is composed of sulfurous acid ($H_2SO_3$), sulfuric acid ($H_2SO_4$), nitrous acid ($HNO_2$), and nitric acid ($HNO_3$), and it is formed from the gaseous products of combustion, specifically by the oxides of sulfur ($SO_2$ and $SO_3$) and nitrogen ($NO$ and $NO_2$). In these experiments, you will observe the simulated environmental effects of acid rain. Your "world" will be a covered petri dish containing drop-sized "lakes," small chunks of "buildings," pieces of metal "bridges," and sources of $SO_2(g)$ and $NO(g)$.

## GENERATION OF SO$_2$ AND NO

Rather than producing these acid rain gases by combustion, you will generate them with the following chemical reactions, which are not representative of how they are formed during combustion:

$$3\,NO_2^-(aq) + 3\,H^+(aq) \rightarrow 3\,HNO_2(aq) \rightarrow 2\,\mathbf{NO}(g) + HNO_3(aq) + H_2O(l)$$

$$SO_3^{-2}(aq) + 2\,H^+(aq) \rightarrow \mathbf{SO_2}(g) + H_2O(l)$$

## RELEVANT REACTIONS OF SO$_2$ AND NO

The $SO_2$ and $NO$ you produce will undergo some of the same reactions they would in the atmosphere. The relevant reactions for this lab are as follows: (a) $SO_2$ reacts with water to yield sulfurous acid; (b) aqueous sulfurous acid dissociates into $2\,H^+$ and $SO_3^{2-}$ ions; (c) nitric oxide reacts with oxygen to form nitrogen dioxide; (d) $NO_2$ reacts with water to form nitrous acid and nitric acid; and (e) aqueous nitrous acid and nitric acid dissociate to form $H^+$, $NO_3^-$ and $NO_2^-$ ions.

a) $SO_2(g) + H_2O(l) \rightarrow H_2SO_3(aq)$

b) $H_2SO_3(aq) \rightarrow 2\,H^+(aq) + SO_3^{2-}(aq)$

c) $2\,NO(g) + O_2(g) \rightarrow 2\,NO_2(g)$

d) $2\,NO_2(g) + H_2O(l) \rightarrow HNO_2(aq) + HNO_3(aq)$

e) $HNO_3(aq) \rightarrow H^+(aq) + NO_3^-(aq); HNO_2(aq) \rightarrow H^+(aq) + NO_2^-(aq)$

Finally, there is a relevant multistep reaction (f), wherein $SO_2$ is oxidized by $NO_2$ to form $SO_3$ and NO. The $SO_3$ then reacts with water to yield sulfuric acid, which dissociates into $2\,H^+$ and $SO_4^{2-}$ ions.

f) $SO_2(g) + NO_2(g) \rightarrow SO_3(g) + NO(g)$

$\phantom{f)}\ SO_3(g) + H_2O(l) \rightarrow H_2SO_4(aq)$

$\phantom{f)}\ H_2SO_4(aq) \rightarrow 2\,H^+(aq) + SO_4^{2-}(aq)$

In your petri dishes, the "lakes," "buildings," or "bridges" placed at various distances from the "pollution source" will absorb some of the $SO_2$ and $NO_x$. The presence of these gases will be detected by probes—chemicals that react with $SO_2$, NO, $NO_2$, or their products. *Your task is to design experiments to determine what compounds are formed when NO and $SO_2$ are emitted.*

## PROBES

- Bromocresol green is an acid–base indicator. It is yellow in acidic solutions and blue in basic solutions.

- Potassium iodide (KI) detects the presence of $NO_2$. Aqueous solutions of KI are colorless, but in the presence an oxidizing agent such as $NO_2$, the iodide ion ($I^-$) is oxidized to elemental $I_2$, which appears brown.

$$2\,NO_2(g) + 2\,I^-(aq) \rightarrow 2\,NO_2^-(aq) + I_2(aq)$$

- Barium chloride ($BaCl_2$) serves as a probe for the sulfate ion ($SO_4^{2-}$), which is detected when sulfuric acid is present. An insoluble white precipitate of barium sulfate ($BaSO_4$) is formed when the sulfate ion is present in solution.

$$SO_4^{2-}(aq) + Ba^{2+}(aq) \rightarrow BaSO_4(s)$$

**Safety Precautions**

Please be careful! Ingested or inhaled sulfites can cause allergic reactions. $SO_2$ and $NO_x$ gases are toxic, so keep the petri dish covered. Dilute with excess water to stop production of $SO_2$ and $NO_x$ gases.

## PROCEDURE

You will be designing experiments that allow you to observe the transport of $SO_2$ and $NO_x$ as well as their effects on lakes of various sizes and distances from the "pollution source."

- To generate $SO_2$ gas, add a drop of sulfuric acid ($H_2SO_4$, the "initiator") to a drop of sodium sulfite ($Na_2SO_3$, the "$SO_2$ source").

■ To generate NO gas, add a drop of $H_2SO_4$ (again, the "initiator") to a drop of sodium nitrite ($NaNO_2$, the "NO source").

■ Once you initiate the reaction, cover the petri dish immediately to contain the gases.

■ Stop the reaction by flooding the petri dish with water.

For each experiment, plan the placement of the drops in the petri dish carefully. Ideally, only one gas source and effect should be tested at a time. You will have a pH probe (bromocresol green), an $NO_2$ probe (KI), and an $SO_4^{2-}$ probe ($BaCl_2$). Think about whether you wish to place your probes at the same distance from the source or at a variety of distances. You may wish to use the premeasured grids in the following figure under your petri dish to accurately place your probes. Some reactions can be observed best over a black background and some over a white background. When you are timing changes, decide whether you want to measure when the change starts or when the change is complete.

Make sure you sketch your experimental setup and record your measured distances and reaction times in your laboratory notebook for *each trial*.

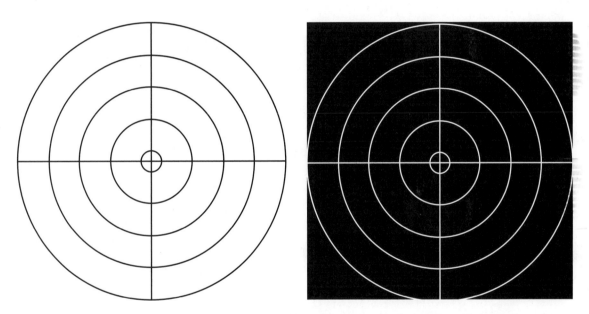

Grids for use under a petri dish. Consider which probe will be used, that is, what is the expected observation, when selecting which grid to use.

1. Does $SO_2$ or NO affect lakes?
   • Study both $NaNO_2$ and $Na_2SO_3$ as the source. Be sure to use only *one source at a time*.
   • Use $H_2SO_4$ as the initiator.
   • Use bromocresol green solution, the acid–base probe, as the "lakes."

a) Does $SO_2$ or NO gas move through the air? How can you tell?

b) Does $SO_2$ or NO have any effect on the lakes? How can you tell?

c) How does the time to observe effects vary for equal-sized lakes at different distances from the source?

d) How does the time to observe effects vary for different-sized lakes at equal distances from the source?

e) Do both $SO_2$ and NO give the same results?

f) What reaction(s) must have occurred to explain your observations?

2. Is $NO_2$ produced?
   - Use $NaNO_2$ as the source.
   - Use $H_2SO_4$ as the initiator.
   - Use the redox probe (KI) as the "lakes."

a) Does NO or $NO_2$ gas move through the air? How can you tell?

b) Does NO or $NO_2$ have any effect on the lakes? How can you tell?

c) The source plus initiator makes NO(g). The probe tests for $NO_2$. Where does $NO_2$ come from? What reaction(s) must have occurred to explain your observations?

3. Is $SO_4^{2-}$ produced?
   - Use $Na_2SO_3$ as the source.
   - Use $H_2SO_4$ as the initiator.
   - Use the $SO_4^{2-}$ probe ($BaCl_2$) as "lakes."

a) Does $SO_2$ gas move through the air? How can you tell?

b) Does $SO_2$ have any effect on the lakes? How can you tell?

c) What if your world were to have both cars and power plants? Use $NaNO_2$ solution as one source and $Na_2SO_3$ solution as a second source. The figure below shows a possible arrangement of sources and probe. Use $H_2SO_4$ solution as the initiator for both. Use $BaCl_2$ solution as an $SO_4^{2-}$ probe in the "lake." Does using both NO and $SO_2$ sources change your answers when using the $SO_4^{2-}$ probe?

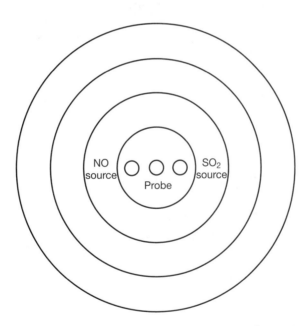

Suggested placement of "cars" and "power plants" around a probe.

d) What reaction(s) must have occurred to explain your observations?

ChemConnections Activity Workbook

4. Will $SO_2$ affect buildings and bridges?
   - Use $Na_2SO_3$ as the source.
   - Use $H_2SO_4$ as the initiator.
   - Place marble chips or small piles of powdered $CaCO_3$ and Mg in the petri dish. Add a drop of bromocresol green onto each sample to serve as the probe.
   - Three minutes after starting $SO_2$ generation, remove the "pollution source" with a cleaning tissue, replace the cover and continue to observe. Flood the cleaning tissue with water to stop gas generation before discarding it.

a) What do you observe?

b) The reaction of the building is

$$H_2SO_4(aq) + CaCO_3(s) \rightarrow CaSO_4(s) + CO_2(g) + H_2O(l)$$

$CaSO_4$ occupies more space than $CaCO_3$. How will acid rain affect the structural integrity of marble objects?

c) The reaction of the bridge is

$$H_2SO_4(aq) + Mg(s) \rightarrow MgSO_4(s) + H_2(g)$$

How will acid rain affect the structural integrity of metallic objects?

d) Is $CaCO_3$ acidic or basic? What do you observe before $SO_2$ is generated and after the $SO_2$ generation is stopped? What reaction(s) must have occurred to explain your observations?

5. Observe the effects of $NO_x$ on lakes from different parts of the country.
   - Use $NaNO_2$ or $Na_2SO_3$ as the source.
   - Use $H_2SO_4$ as the initiator.
   - Use small portions of the solutions you prepare as the lakes.

In this experiment, you will prepare solutions with varying concentrations of sodium bicarbonate, $NaHCO_3$, in a spot plate or separate test tubes. These solutions represent lakes from different geographic regions that are exposed to different amounts of limestone.

| NaHCO$_3$ Solution | Bromocresol Green Solution |
| --- | --- |
| 1 drop | 9 drops |
| 2 drops | 8 drops |
| 3 drops | 7 drops |
| 4 drops | 6 drops |

Use drops of these prepared solutions to represent lakes from regions that have different geology.

   a) Keep your lakes small and all the same size and the same distance from the source. How does time vary to observe effects with lakes made from each mixture? Include a graph of time as a function of the relative $NaHCO_3$ concentration in your laboratory notebook.

   b) Lakes in regions with abundant limestone have relatively high concentrations of $NaHCO_3$ compared to lakes in granitic regions. Do you think the effects of acid rain would be observed more rapidly in lakes exposed to limestone or those exposed to granite? Explain.

6. If you were designing an environmental study to test the effects of acid rain on a lake, would you select a small or large lake? Explain.

Adapted by Kevin Braun, Brock Spencer, and George Lisensky from the ChemConnections module "Soil Equilibrium: What Happens to Acid Rain?" by George Lisensky, Roxanne Hulet, Michael Beug, and Sharon Anthony.

# What Is the Mass of Air?

- To explore the functional relationships among pressure, temperature, volume, and mass for gases

## INTRODUCTION

Why does the National Football League require all footballs to be filled to between 12.5 and 13.5 psi? Why does your car manual suggest checking the tire pressure before driving? Why are weather balloons substantially underfilled with helium when they are prepared for flight? In this activity, you will explore the relationships among volume, temperature, pressure, and mass for gases to help answer such questions.

## PART I: WHAT IS THE DENSITY OF AIR?

Many sports, including football, soccer, and basketball, mandate the pressure of the balls they use. Beyond the need to have a consistent bounce, there is another reason that pressure in the ball is critical. You can determine this reason with the results of a simple experiment in which a 2 L bottle is pressurized and then the air is incrementally released. At each pressure, as measured with a tire gauge, the mass of the bottle is recorded until the pressure reaches zero. Sample experimental data are provided in the table below and should be used to answer the questions in Part I.

| Bottle Pressure (psi) | Bottle Mass (g) |
|---|---|
| 40 | 69.22 |
| 30 | 67.75 |
| 20 | 66.08 |
| 10 | 64.35 |
| 0 | 62.64 |

Experiment conducted at 25°C

1. What variables (pressure, temperature, mass, or volume) are kept constant in this experiment? What variables are changing?

2. Use a spreadsheet program or a graphing calculator to fit a trend line to the data ($x$ = pressure; $y$ = mass). Write the equation of the line here.

3. When the tire gauge reads zero there is actually 1 atm of gas pressure in the bottle. Using the trend line, calculate the weight of the bottle if all the air were removed (pressure of $-14.7$ psi).

4. What is the mass of air left in the bottle when the tire pressure gauge reads zero?

5. What is the functional relationship between the variables that changed? If one variable increases, what does the other do?

6. What is the density of air at 25°C? Density is the mass per unit volume of a substance.

7. Based on your previous answer, why does the National Football League, along with governing bodies of other sports, regulate the pressure of the balls used in competition?

## PART II: WHEN SHOULD I MEASURE THE PRESSURE IN MY TIRES?

Why do car manuals suggest tire pressure be measured before driving or after extreme temperature shifts? The answer to this can be found by monitoring the volume of air in a 250 mL flask at varying temperatures. A 250 mL Erlenmeyer flask fitted with a one-hole stopper is placed in a beaker of boiling water (100°C). What happens to the air in the flask as it heats up? Once the flask has acclimated to the boiling water, the stopper hole is covered and the flask is inverted and placed in a bucket of room temperature water (25°C). Upon acclimation, the stopper hole is uncovered, causing a small volume of water to rush into the flask. What causes the vacuum that draws the water into the flask? The air pressure in the flask is balanced by matching the water level inside the flask with that in the water bucket. The stopper hole is then covered and the flask removed from the bucket. The volume of water in the flask is measured using a graduated cylinder. The experiment is repeated at two additional temperatures and the data are summarized here. Use the data to answer the following questions.

| Temperature (°C) | Water Volume (mL) |
|---|---|
| 100 | 0.0 |
| 50 | 36.3 |
| 25 | 53.9 |
| 0 | 72.8 |

8. The total volume of the flask is 273.1 mL as determined by water displacement. Why is the total volume not 250 mL? Use the total flask volume and the data in the preceding table to determine the air volume at each temperature. Summarize your answers in the following table.

| Temperature (°C) | Air Volume (mL) |
|---|---|
| 100 | 273.1 |
| 50 | |
| 25 | |
| 0 | |

9. What variables (pressure, temperature, mass, or volume) are kept constant in this experiment? What variables are changing?

10. What is the functional relationship between the two variables that changed?

11. Use a spreadsheet program or a graphing calculator to fit a trend line to the data ($x$ = temperature; $y$ = air volume). Write the equation of the line below.

12. Use the trend line to determine the temperature required to reduce the volume of air to zero. What is the special name for this temperature? Why is the experimentally found temperature not exact?

13. Based on your previous answers, why do car manuals suggest tire pressure be measured before driving or after extreme temperature shifts?

## PART III: WHY ARE WEATHER BALLOONS UNDERFILLED UPON LAUNCH?

Weather balloons play a critical role in our ability to predict weather. When launched, weather balloons are significantly underfilled. Why is this? The reason can be elucidated by a simple experiment using a syringe and a bathroom scale. The syringe is filled with 60 mL of air and the end is sealed. The air in the syringe is then compressed by pressing down on the plunger with the sealed syringe tip placed on a bathroom scale. The more the air in the syringe is compressed, the larger the weight registered by the scale. At successive air volumes, the scale weight is recorded. Use the data table here to answer the following questions.

| Air Volume (mL) | Scale Weight (lb) |
|---|---|
| 60 | 0.0 |
| 50 | 2.2 |
| 40 | 5.2 |
| 30 | 10.4 |
| 20 | 19.3 |
| 10 | 42.4 |

14. The scale registers in pounds (not pressure). The mass reading can be converted to pressure by first determining the cross-sectional area of the syringe plunger. If the diameter of the syringe pump is 1 in, what is the cross-sectional area of the plunger?

15. Use the plunger area plus the scale reading to convert the scale readings to relative pressure in pounds per square inch (psi). When the syringe is capped, the pressure inside is actually one atmosphere. Correct your relative pressure readings to absolute pressure by adding 14.7 psi to each reading. Summarize your answers in the table here.

| Air Volume (mL) | Relative Pressure (psi) | Absolute Pressure (psi) |
|---|---|---|
| 60 | | |
| 50 | | |
| 40 | | |
| 30 | | |
| 20 | | |
| 10 | | |

16. What variables (pressure, temperature, mass, or volume) are kept constant in this experiment? What variables are changing?

17. Use a spreadsheet program or a graphing calculator to plot the data ($x$ = pressure; $y$ = air volume). Also graph $x$ = pressure and $y$ = 1/volume. Based on these plots, what is the functional relationship between the two variables that changed?

18. Why are weather balloons significantly underfilled when launched? What would happen if they were completely filled before launch?

Written by Kevin Braun, Brock Spencer, and George Lisensky.

# How Do the Gas Laws Affect My Outdoor Experience?

- **To perform calculations using the ideal gas law**

- **To apply the gas laws to everyday situations**

## INTRODUCTION

When studying the gas laws, it's easy to wonder whether there are any real-world applications. In fact, the gas laws have a wide range of applications in your everyday activities, from biking to rafting. The following problems highlight some of these examples.

## PROBLEMS

1. River rafters often fill their rafts in the chill of the morning. While rafting down the cool river, their rafts stay cool. However, when they pull out for lunch and take their rafts partway out of the river, their rafts can warm up quickly. If a raft warms up too much, it can expand to the point of explosion. Needless to say, a raft with an exploded tube is not very useful in the middle of a wilderness river trip.

   a) What can rafters do to prevent raft explosions?

   b) Assume that the raft was filled in the morning when it was 50°F. The initial volume of one of the raft's tubes was 19 ft³. At lunch on a Grand Canyon expedition, the raft's tube popped when it reached 110°F. What was the volume of air, in ft³, when the tube popped?

2. A camp stove's propane canister holds approximately 1 lb of propane ($C_3H_8$) fuel. Assume the fuel canister started at 70°F. Will the temperature of the canister increase or decrease as fuel is burned? Explain.

3. A cyclist fills her tire to a pressure of 100 pounds per square inch (psi) at night when it is 90°F. In the morning, her tire pressure gauge reads only 85 psi. What is the morning temperature? (*Hint:* 1 atm = 14.7 psi)

4. A bicycle tire holds approximately 2.0 L of air. Assume that the temperature is 25°C and the tire pressure is 100 psi.

   a) How many moles of air are in the tire?

   b) How many molecules of air are in the tire?

5. A certain bike tire will burst if it exceeds 130 psi. A cyclist filled it to 100 psi in the winter in northern Wisconsin when it was −20°F. On a 90°F summer day, the cyclist decides to go for a bike ride. When she retrieves her bike from the depths of the garage, will her tire be popped?

6. A basketball has a volume of roughly 412 in³. Calculate how many molecules of air there are per cubic centimeter in the basketball, assuming the temperature is 25°C and the pressure is 1 atm.

7. The tires are low on a dune buggy. On going to the gas station to top the tires off, the owner meets a friend who suggests that he fill his tires with nitrogen rather than air because it will make the dune buggy lighter and thus faster.

   a) The average dune buggy tire has a volume of approximately 10 L. Assume 26.67°C and 30 psi. How many moles of air are required to fill the four tires?

   b) Assuming the density of air under these conditions is 0.224 lb/ft³, calculate the mass of air required to fill the four tires.

   c) What is the mass of nitrogen required to fill the four tires to 30 psi at 26.67°C?

   d) Should the owner pay the extra cost to fill his tires with nitrogen?

# How Can We Predict the Light Produced by Hot Objects?

## INTRODUCTION

The electromagnetic radiation emitted by heated objects is called blackbody radiation. A blackbody is one that absorbs all incident radiation and whose radiant emission is dependent only on its temperature. (All blackbody materials emit the same color and intensity at a given temperature.) Even cool objects emit such radiation, but in a portion of the electromagnetic spectrum that is invisible to our eyes, namely, the infrared region.

A blackbody radiation curve is a plot of the intensity of light emitted as a function of wavelength. One of the triumphs of modern quantum mechanics is that with Planck's assumption, the shapes of the blackbody radiation curves can be quantitatively predicted. This exploration examines the shapes of these theoretical curves.

## PART I: BLACKBODY RADIATION

Use either the graphs provided here or the interactive color versions at www.wwnorton .com/college/chemistry/chemconnections/BlueLight/pages/body.html to explore how color (wavelength) and intensity are related to the temperature of an object.

1. How does the *intensity* for a blackbody radiator vary qualitatively with the temperature?

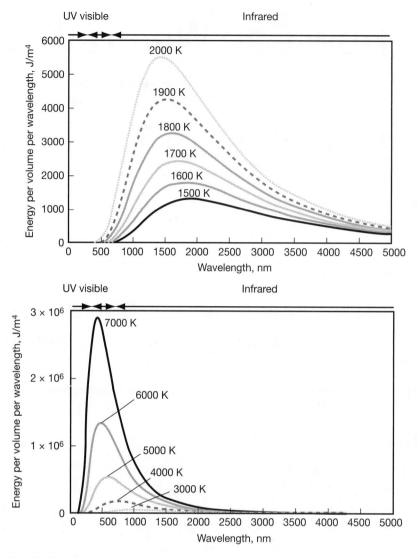

Blackbody radiation curves for various temperatures

2. How does the *color (wavelength) distribution* for a blackbody radiator vary qualitatively with the temperature?

3. Examine the blackbody radiation curves.

   a) What temperature would be required to emit primarily *red* light?

   b) What temperature would be required to emit primarily *green* light?

   c) What temperature would be required to emit primarily *blue* light?

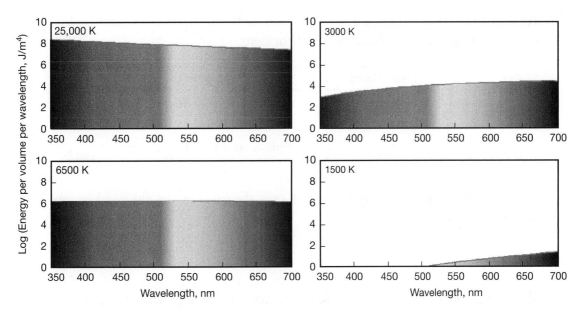

Visible portion of blackbody radiation log curve (only a small range of wavelengths can be seen by eye)
A color version of this figure can be found at www.wwnorton.com/college/chemistry/chemconnections/BlueLight/pages/body.html

## PART II: APPLICATIONS

Apply your knowledge from Part I to the following examples.

4. Look at the following figure and determine which is hotter: a flame from methane and air (left) or a flame from methane and oxygen (right)? How could you tell from these pictures?

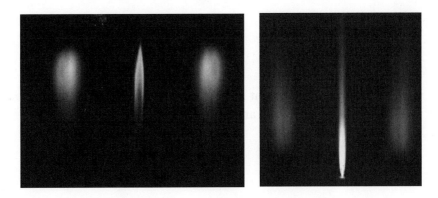

Flame viewed through a diffraction grating
A color version of this figure can be found at www.wwnorton.com/college/chemistry/chemconnections/BlueLight/pages/grate.html

5. At relatively low pressures, a number of gases will emit light of a specific color if an electric charge is passed through them. For example:

| Gas | Color of Light | Practical Application |
| --- | --- | --- |
| Mercury | Blue | Mercury vapor lights for parking lots |
| Neon | Red | Neon lights for signs |
| Sodium | Yellow | Sodium vapor lights for highways |

Based on the frequency of light emitted, rank the gases by the electrical energy required to emit light.

6. Blackbody radiation curves for different temperatures are given in the following figure. The maximum in the curve corresponds to the wavelength of maximum intensity. Identify which is white hot and which is red hot, and which is at the higher temperature. Note that the $x$- and $y$-axis scales are different in the two graphs.

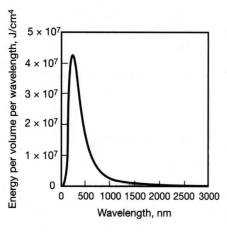
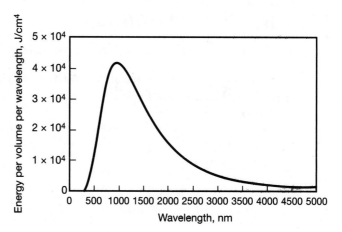

Blackbody radiation curves

7. When you turn off an incandescent lightbulb (stop the electrical flow) does the light go out immediately? Explain your answer.

8. Stars are not *solids*, but they are opaque and do exhibit blackbody radiation.

### Properties of Various Classes of Stars

| Star Class | Approximate Temperature (K) | Color |
|---|---|---|
| O | 31,000 | Blue-white |
| B | 20,500 | Blue-white |
| A | 11,500 | White |
| F | 8,000 | Yellow-white |
| G | 6,500 | Yellow |
| K | 4,500 | Orange |
| M | 3,000 | Red |

a) Estimate the surface temperature of the Sun.

b) You have seen objects glow "red hot." At higher temperatures, objects are "orange hot" and then "yellow hot" and then "white hot." Some stars are "blue hot." Estimate what temperature would be required for a substance to glow "blue hot."

9. The following dialog box allows an operator to set the "white point" for a computer monitor. Why are the units for white in Kelvin? Explain why the values do or do not make sense.

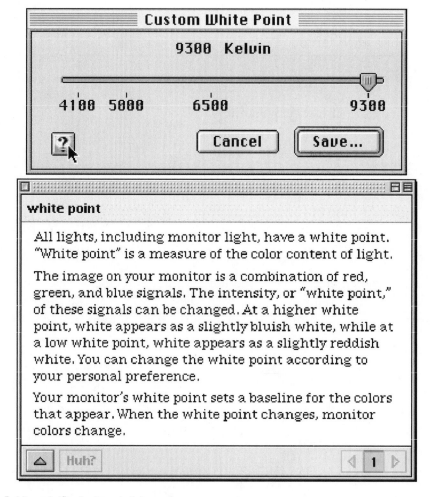

Computer "white point" adjustment dialogue box.

10. The Sun and the Earth both exhibit blackbody emissions, as seen in the figure below. Which curve corresponds to which? Explain your answer by discussing the emission temperature, wavelength, and intensity.

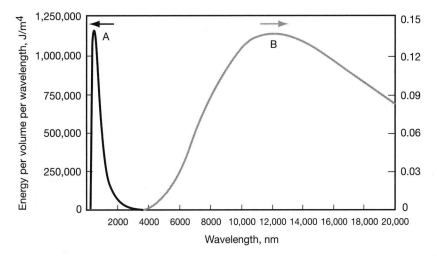

Blackbody emission for two different temperatures. Curve A corresponds to the scale at the left; curve B corresponds to the scale at the right.

Adapted by George Lisensky from the ChemConnections module "Build a Better CD Player: How Can You Get Blue Light from a Solid?" by George C. Lisensky, Herbert Beall, Arthur B. Ellis, Dean J. Campbell, and Joanne Stewart.

ACTIVITY

25

# Can We Identify Unknown Plastics Using Infrared Spectroscopy?

## LEARNING GOALS

■ **To become familiar with infrared (IR) spectroscopy**

■ **To relate IR spectra to the bonds present in different polymers**

■ **To identify polymers based on their IR spectra**

## INTRODUCTION

Infrared spectroscopy is based on the premise that when light of a particular frequency is directed at a molecule, it may stimulate the vibration of a particular type of bond within the molecule. If this occurs, that particular frequency of light will be absorbed and the amount of absorption can be measured using an infrared spectrophotometer. The frequency of light that is absorbed depends on two factors: the masses of the atoms involved and the relative stiffness of the bonds. Thus, by determining the frequency of light absorbed by the molecule, we can determine the types of bonds in a molecule. Infrared spectra are typically plotted as percent transmittance on the *y*-axis and wavenumbers on the *x*-axis. (A wavenumber is the inverse of the wavelength of light involved, and thus is related to the frequency of the light.)

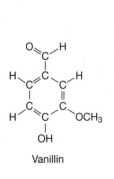

Vanillin

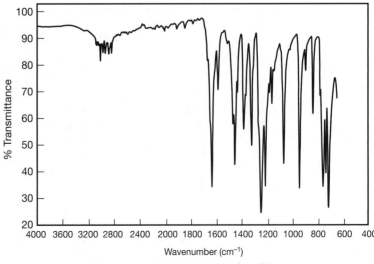

Infrared spectrum of vanillin

**109**

As an example, the infrared spectrum of vanillin is shown above. In regions where the percent transmittance is high, light is not absorbed. In regions where light is absorbed, there are valleys in the spectrum. The deeper the valley, the greater the absorption. The valleys in an infrared spectrum are often referred to as "peaks."

As described previously, certain portions of the IR spectrum are associated with particular molecular vibrations. The table here gives some specific examples of the portion of the spectrum associated with certain types of bonds. For example, because the energy absorbed by a $C=C$ bond (~1650 cm$^{-1}$) is different from that absorbed by $C \equiv C$ bond (~2175 cm$^{-1}$), it is possible to use infrared spectroscopy to determine which type of bond is present in a molecule. In this laboratory data analysis exercise, you will use infrared spectroscopy to identify the bonds present in a variety of different molecules including plastics.

# IMPORTANT INFRARED BANDS

| TYPE OF BOND | WAVENUMBER (cm⁻¹) | INTENSITY |
|---|---|---|
| N≡C | 2260–2220 | Medium |
| C≡C | 2260–2100 | Medium |
| C=C | 1680–1600 | Medium |
| Benzene | ~1230 and ~1500 | Strong |
| C=O | 1780–1650 | Strong |
| C—O | 1250–1050 | Strong |
| C—N | 1230–1020 | Medium |
| O—H (alcohol) | 3650–3200 | Strong, broad |
| O—H (acid) | 3300–2500 | Strong, very broad |
| N—H | 3500–3300 | Medium, broad |
| C—H | 3300–2700 | Medium |

1. Polystyrene is a polymer with the following structural formula and infrared spectrum:

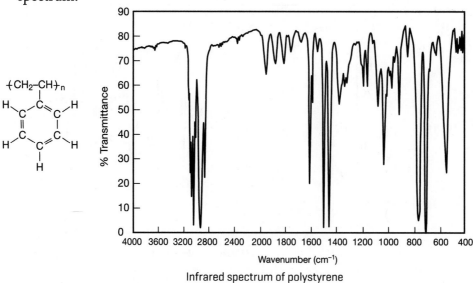

Infrared spectrum of polystyrene

a) Which peak(s) in the spectrum of polystyrene correspond to the benzene ring in polystyrene?

b) What bond(s) cause the peaks near wavenumber 3000 cm$^{-1}$?

2. The structures of benzene and cyclohexane are shown here.

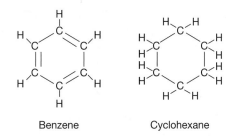

Benzene          Cyclohexane

a) What peak(s) would you expect benzene and cyclohexane to have in common?

b) How would you expect the spectra of benzene and cyclohexane to differ? Give specific peak wavenumbers in your response.

3. Ethylene ($H_2C{=}CH_2$) is the monomer used to produce polyethylene, $-(H_2C{-}CH_2)_n$

a) Would you expect the major peak(s) in the spectrum of ethylene to be the same as those in polyethylene? Why or why not?

b) The spectrum of either ethylene or polyethylene is given here. Which is it? Explain your reasoning.

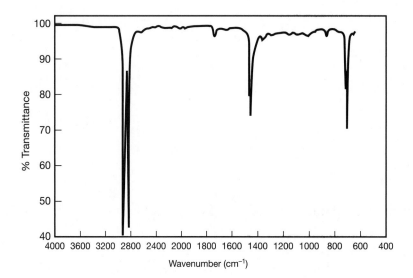

4. Where would you expect to find peaks in the spectrum of the following molecules?

a) Styrene

b) Acetic acid

5. Three IR spectra appear here. They are for the compounds:

Methyl ethyl ketone, CH$_3$COCH$_2$CH$_3$

1-Butanol, CH$_3$CH$_2$CH$_2$CH$_2$OH

Diethyl ether, CH$_3$CH$_2$OCH$_2$CH$_3$

Draw the structural formula for the proper compound next to the corresponding infrared spectrum, and label the major peaks with the type of bond being stretched.

a)

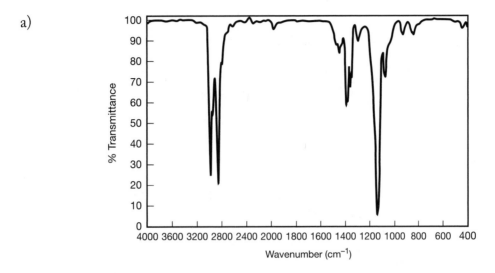

b)

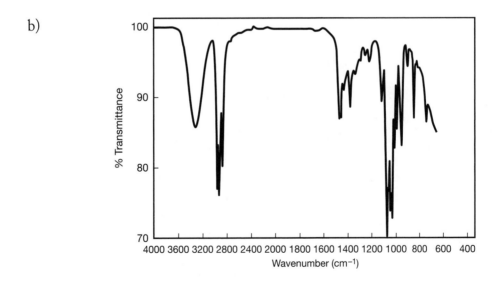

c)

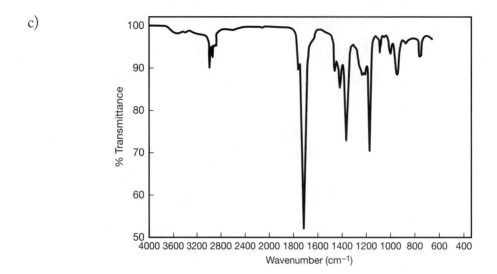

Adapted by Sharon Anthony, Karen Harding, and Kim Kostka from the ChemConnections module "How Do We Get from Bonds to Bags, Bottles, and Backpacks?" by Karen Harding and Sharon Anthony.

# What Is a Greenhouse Gas?

## INTRODUCTION

Sunlight that reaches the Earth consists of a rainbow of visible light, plus invisible infrared (IR) radiation and ultraviolet (UV) radiation. Sunlight that is absorbed by the Earth is then radiated back to the atmosphere in the form of IR radiation, which your body perceives as heat. The greenhouse gases—such as carbon dioxide ($CO_2$), methane ($CH_4$), water vapor ($H_2O$), refrigerants such as 1,1,1,2-tetrafluoroethane ($CH_2FCF_3$, or HFC-134a), and nitrous oxide ($N_2O$)—trap much of the heat that is radiated from the surface of the Earth. Air is composed predominantly of nitrogen ($N_2$) and oxygen ($O_2$), which fortunately do not behave as greenhouse gases. The purpose of this activity is to identify the properties that are unique to greenhouse gases and to understand why some gases are greenhouse gases whereas others are not.

Some of the unique properties of greenhouse gases can be determined by using spectroscopy to investigate the interaction of atmospheric gases with visible and infrared light. You will look at the infrared and visible absorption characteristics of $N_2$, $O_2$, $CO_2$, $CH_4$, $H_2O$, $CH_2FCF_3$, and $N_2O$. From these results, you will be able to deduce the absorption features that make greenhouse gases unique in our atmosphere. This is a discovery exercise; therefore, do not be frustrated if you do not know what you are looking for at the onset.

# ANALYZING INFRARED SPECTRA

Use the following spectra to answer questions 1–3.

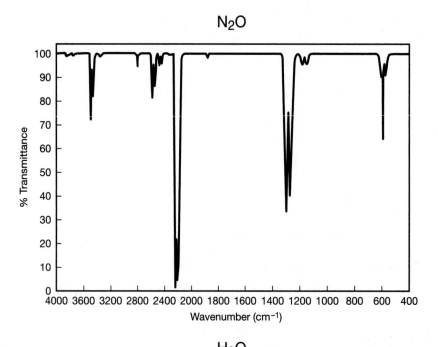

N₂O

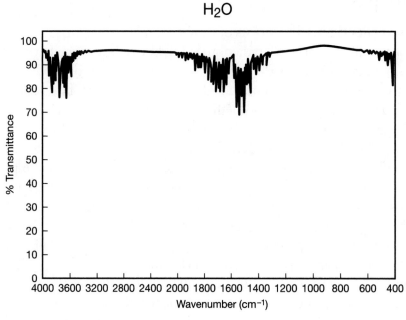

H₂O

## CH₄

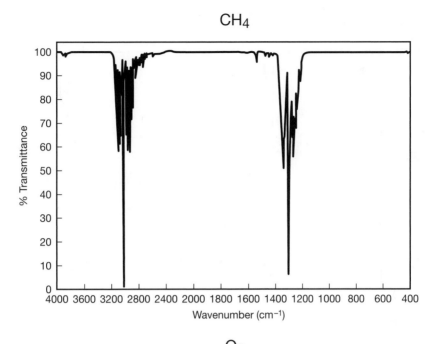

## O₂

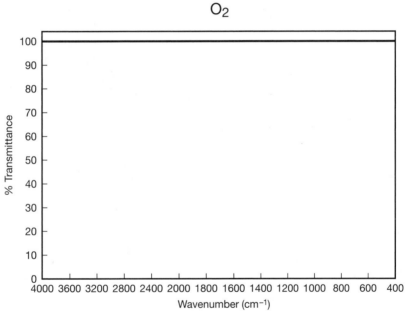

*(continued)*

## CO₂

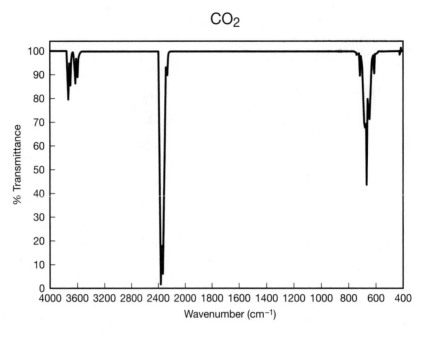

## CH₂FCF₃

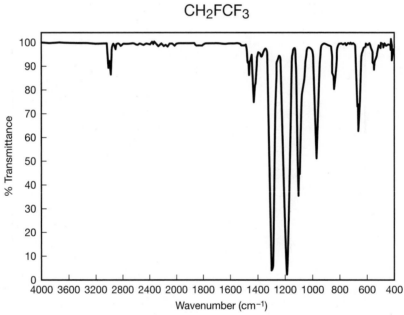

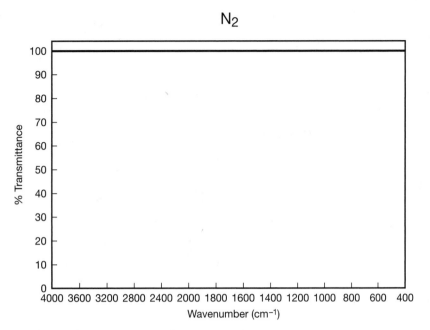

Infrared spectra of gases found in the atmosphere.

1. What are the wavenumber ranges where each gas absorbs infrared light? (Note at least three major peaks, if applicable.)

| Molecule | Wavenumber Range (cm$^{-1}$) | Wavenumber Range (cm$^{-1}$) | Wavenumber Range (cm$^{-1}$) |
|---|---|---|---|
| $CO_2$ | | | |
| $O_2$ | | | |
| $N_2$ | | | |
| $CH_4$ | | | |
| $N_2O$ | | | |
| $CH_2FCF_3$ | | | |
| $H_2O$ | | | |

2. Circle the gases that strongly absorb infrared light.

$$O_2, N_2, CO_2, CH_4, H_2O, N_2O, CH_2FCF_3$$

3. Circle the gases that do not strongly absorb infrared light.

$$O_2, N_2, CO_2, CH_4, H_2O, N_2O, CH_2FCF_3$$

4. Based on the following picture of glass tubes containing the gases, circle the gases that strongly absorb visible light.

$$O_2, N_2, CO_2, CH_4, H_2O, N_2O, CH_2FCF_3$$

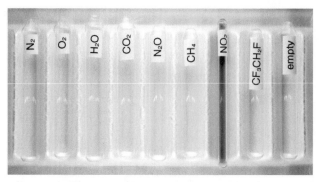

A color version of this figure can be obtained from your instructor.

5. Spectroscopically, how do greenhouse gases differ from nongreenhouse gases?

6. Draw a sketch of the greenhouse effect that includes the Sun, the atmosphere, the Earth, visible light, and infrared light.

7. Your breath is a mixture of gases. Analyze the spectrum below, and identify the gases that can be detected in your breath using infrared spectroscopy.

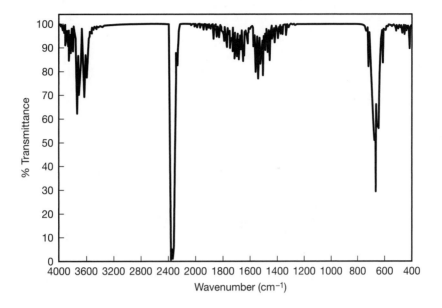

Adapted by Brock Spencer and Kevin Braun from the ChemConnections module "What Should We Do about Global Warming?" by Sharon Anthony, Tricia Ferrett, and Jade Bender.

# Laboratory: Why Do We Need Ozone?

- **To use ultraviolet-visible spectroscopy to understand the bioprotective role of ozone**

- **To gain experience in experimental design**

## INTRODUCTION

Why worry about the ozone hole? What does ozone do to protect our existence? As you may have heard, ozone is often described as the "Earth's sunscreen" because it shields us from the damaging ultraviolet (UV) rays that cause our skin to tan or burn. In addition, overexposure to UV radiation has been linked to a variety of adverse health effects, from increased skin cancer and cataracts to mutation of DNA. In fact, life on the surface of the planet Earth would be impossible without ozone.

In this activity, you will perform two experiments to help you understand how atmospheric gases and sunscreens absorb UV radiation from the Sun and thus protect us from the harmful effects of the Sun. The first experiment on ultraviolet-visible (UV-Vis) absorption spectroscopy allows you to measure the wavelengths at which a substance absorbs UV light. You will have a chance to do this for the only important UV absorber in the atmosphere, ozone, and for sunscreens. In a second experiment, you will use some colorful UV beads as detectors of solar UV radiation. You will be asked to design an experiment to assess the ability of several materials to absorb and thus block UV light.

## PART I: SPECTROSCOPY EXPERIMENTS

You will measure and record UV spectra of atmospheric gases and sunscreens. Teams will share data so everyone can see spectra of all available substances, which may include ozone ($O_3$), oxygen ($O_2$), a chlorofluorocarbon (CFC) or a CFC-replacement gas, and sunscreens with sun protection factors (SPFs) of 4, 8, 15, 30, and 45. Listen for instructions about which gases and sunscreens your team will investigate.

ChemConnections Activity Workbook

## UV Spectra of Atmospheric Gases

The visible part of the spectrum, defined by what the human eye detects, is from about 400 to 700 nm. The higher-energy UV range, which causes sunburn and damage to biological systems, is below 400 nm. The UV spectrum is often divided into three separate regions: UV-A (320–400 nm; least energetic), UV-B (290–320 nm), and UV-C (200–290 nm; most energetic).

For this analysis, you will need spectra for $O_3$, $O_2$, a CFC, or a CFC-replacement gas from 190 to 1100 nm. You will either prepare the spectra you will use or use spectra that have already been recorded. Ozone is prepared by passing an electric current through oxygen. The electric current is energetic enough to break the oxygen–oxygen double bond to form two oxygen (O) atoms, each of which will quickly recombine with an $O_2$ molecule to produce $O_3$. Your instructor will either assist you in preparing an ozone sample or provide the spectrum of ozone for you to work with.

## Sunscreen Spectra

Your instructor will either provide you with spectra that have been recorded previously or help you run UV-Vis spectra of *one* of the following sunscreens: SPF 4, 8, 15, 30, or 45. Coordinate with other groups so that all SPF sunscreens get analyzed. To take a spectrum of sunscreen, use the quartz cell with air as a blank. To take a sunscreen sample spectrum, *gently* smear a very thin, translucent layer of sunscreen on the inside of one clear face of the quartz cell, using a soft foam swab. If the absorbance is greater than 1.0, remove some sunscreen and try again; repeat if necessary. Immediately after use, gently clean the sunscreen off the quartz cuvette with isopropanol and a cotton swab.

The active ingredients in sunscreens change over time as manufacturers improve their products. A sample active ingredient list for some representative sunscreens follows, along with the corresponding chemical structures. Other active ingredients you may encounter include octyl methoxycinnamate, avobenzone, octocrylene, octinoxate, octyl salicylate, homosalate, and octisalate.

| SPF | Active Ingredients |
| --- | --- |
| 4 | Ethylhexyl *p*-methoxycinnamate, oxybenzone |
| 8 | Ethylhexyl *p*-methoxycinnamate, oxybenzone |
| 15 | Ethylhexyl *p*-methoxycinnamate, oxybenzone |
| 30 | Ethylhexyl *p*-methoxycinnamate, oxybenzone, 2-ethylhexyl salicylate, homosalate |
| 45 | Ethylhexyl *p*-methoxycinnamate, oxybenzone, 2-ethylhexyl salicylate, homosalate |

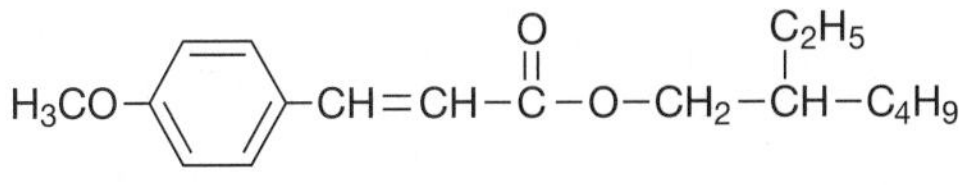

Ethylhexyl p-methoxycinnamate

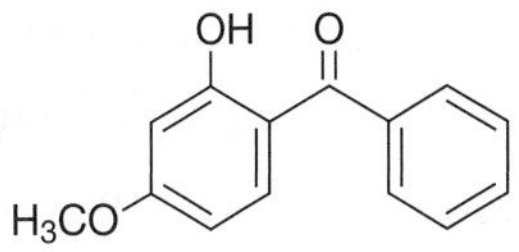

Oxybenzone

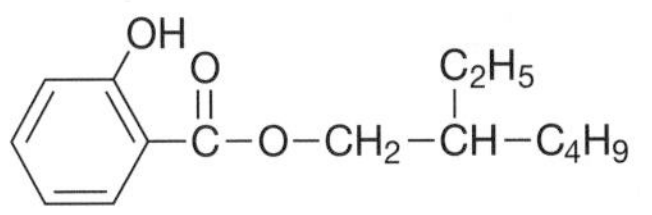

2-ethylhexyl salicylate

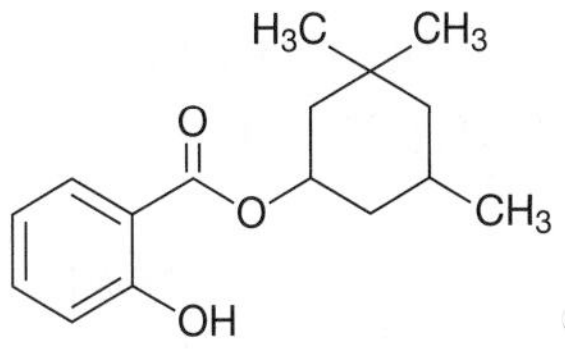

Homosalate

## Ozone Decomposition Rate Study

Can you demonstrate that $O_3$ is an unstable chemical species that participates in interesting chemistry in the stratosphere? To do this, your instructor will set up a cuvette of $O_3$ on a spectrometer and let it record for several hours the UV absorbance ($A$) at the wavelength of maximal absorbance for ozone. What do you expect to see happen over time? Check in once in a while to see the results, and be sure to *check the data before you leave the lab*. Record in words what you observe about the spectral changes with time.

## Data Analysis

1. At what wavelength does $O_3$ absorb the maximum amount of UV light?

2. Does ozone absorb visible light, and how can you tell, using both the spectrum and your eyeballs?

3. Estimate the wavelength range over which $O_2$, a CFC, or CFC-replacement absorb UV light.

4. Use your UV spectra to predict which atmospheric gases have the potential to act as effective sunscreens for Earth. Cite wavelength ranges from your data to support your answer.

5. Compare your UV spectra for $O_3$ with all the sunscreen (SPF 4–45) spectra, noting minima and maxima in each. What do your observations suggest about how manufacturers might be designing sunscreens?

6. Examine the chemical structures for the sunscreens. What do these molecules have in common? Which portion of the molecule do you think is responsible for absorbing UV light?

## PART II: SOLAR EXPERIMENTS

Your instructor will supply you with colorless beads in packets labeled by color. Do not mix up the beads—they look nearly the same (clear) prior to UV exposure. Take care to avoid exposing them to sunlight, even in the lab. These unique beads will turn a variety of colors when exposed to UV radiation from the Sun or from a laboratory UV source. Furthermore, the various colored beads turn from colorless to colored at different rates.

You will need to discuss and design some experiments aimed at finding out which materials protect us from UV radiation. You ought to be able to rank the materials in terms of their protective power. Your mission is to design a series of such experiments using an assortment of items, which may include the supplies listed here.

UV beads, a variety

Bead holder or tray

Two kinds of cut glass of about the same thickness:

  —special glass used on a front windshield of a car, with a layer of plastic inside it
  —regular window glass

Fabric samples

Eyeglasses

Sunglasses, a variety

Sunscreens of SPF 4, 8, 15, 30, and 45

Stopwatch

Foam swabs for applying sunscreen

Large plastic bin

Some kind of bin cover

Natural sunlight

In addition, you will be working under the following constraints:

- Try to use all the items supplied, though some of them may be more or less useful than others.

- You must plan and execute your experiments in the time allotted by your instructor.

First, talk with your group about what experiments you would like to design. Discuss the following:

- What questions are you trying to answer?

- For each question and accompanying experiment you have created, what is your hypothesis for what will happen? Be specific.

When you finish, return the supplies and be prepared to summarize your results.

## Data analysis

7. Describe and draw pictures of your UV-bead experiments.

8. What questions were you trying to answer?

9. Summarize your results and the answers to your proposed questions. Which materials block UV light?

10. Were there any problems with your experiments? Are you unsure of anything you did? What would you do differently next time?

---

Adapted by Sharon Anthony and Kim Kostka from the ChemConnections module "Why Does the Ozone Hole Form?" by Tricia Ferrett and Sharon Anthony. This lab was developed by Melissa Kido and Tricia Ferrett at Carleton College.

---

# What Are the Shapes of the Greenhouse Gas Molecules?

## INTRODUCTION

In this activity, you will draw the Lewis structures of relevant atmospheric gases and then build models of them using toothpicks and gumdrops. Greenhouse gases transmit visible light but absorb infrared light. Carbon dioxide ($CO_2$), methane ($CH_4$), nitrous oxide ($N_2O$), water ($H_2O$), and refrigerants such as trichlorofluoromethane ($CCl_3F$) are greenhouse gases, whereas the prominent atmospheric gases oxygen ($O_2$) and nitrogen ($N_2$) transmit infrared light and, thus, are not greenhouse gases. The molecules sulfur dioxide ($SO_2$) and ammonia ($NH_3$) are included in this activity to give you practice with VSEPR theory, but they are not considered greenhouse gases. Although they do absorb infrared light, they are not present in high enough concentrations in the atmosphere to behave as significant greenhouse gases.

# PART I: STEP-BY-STEP APPROACH FOR DETERMINING MOLECULAR SHAPE

**Step 1:** Draw the Lewis structure for the molecule of interest. Be sure to include all of the valence electrons.

**Step 2:** Count the number of electron domains around a central atom. If there is not a central atom, count the number of electron domains around one of the atoms.

**Step 3:** Determine the *electronic shape* of the molecule.

- Use a gumdrop to represent the central atom.

- Use toothpicks to represent electron domains around the central atom.

- Arrange the toothpicks in the appropriate shape such that the electrons (i.e., toothpicks) are as far apart from one another as possible. Why do you think the electron domains are arranged so far apart?

**Step 4:** Determine the *molecular shape* of the molecule.

- Based on your Lewis structure, place a gumdrop (i.e., atom) on the end of each toothpick (i.e., electron domain) involved in a bond.

- Do not put a gumdrop on any toothpick that does not represent electrons shared between covalently bonded atoms. Toothpicks without gumdrops will represent lone pairs of electrons.

- Now look *only* at the configuration of the atoms (i.e., gumdrops) to determine the molecular shape.

| Molecule | Lewis Structure | Electronic Shape | Molecular Shape |
|---|---|---|---|
| $CH_4$ | | | |
| $O_2$ | | | |
| $H_2O$ | | | |
| $N_2$ | | | |
| $CO_2$ | | | |
| $SO_2$ | | | |
| $NH_3$ | | | |
| $CCl_3F$ | | | |

## PART II: FORMAL CHARGE AND RESONANCE

Sometimes it is possible to draw more than one Lewis structure for a molecule. Formal charge and resonance are tools chemists use to determine which structures are most plausible.

1. Draw all the possible electronic structures for $N_2O$, which has an atomic arrangement of N–N–O and is not cyclic.

2. Assign formal charges to each of the atoms in each of your structures.

3. Based on formal structure, which structure for $N_2O$ is preferred?

4. Use formal charges to predict the structure of $NO_3^-$, an important molecule in atmospheric aerosols.

5. Draw the two possible Lewis structures for $SO_2$. Is one structure preferred over another, based on formal charge?

6. The structures you drew for $N_2O$, $NO_3^-$, and $SO_2$ are called *resonance structures*. Define resonance.

Adapted from the ChemConnections module "What Should We Do about Global Warming?" by Sharon Anthony, Tricia Ferrett, and Jade Bender.

# What Are the Shapes of Molecules in Paper and Plastic Bags?

- To learn to use valence shell electron pair repulsion (VSEPR) theory to predict the shapes of molecules

- To apply valence bond (hybrid orbital) theory to a number of different molecules, including polymers

## INTRODUCTION

Paper or plastic? Have you ever thought about the chemical makeup of your grocery bags? Chemists have developed a number of theories to describe chemical bonding and to predict molecular shapes. Theories of bonding include VSEPR theory and valence bond theory. In this activity, you will use both VSEPR and valence bond theories to predict the shapes of molecules, including the molecules that make up paper and plastic bags.

## VSEPR THEORY

1. Build a model of methane ($CH_4$). Estimate the H–C–H bond angle. Is it greater or less than 90°?

2. Build a model of ethane ($C_2H_6$). Estimate the H–C–C bond angle and the H–C–H bond angle. Are these bond angles the same?

3. Based on VSEPR theory, what is the molecular shape of methane ($CH_4$)? What are the bond angles?

4. Based on VSEPR theory, what is the molecular shape around each carbon atom in ethane $(C_2H_6)$?

5. Build a model of ethylene $(C_2H_4)$. Use VSEPR theory to estimate the $C=C-H$ bond angle and the $H-C-H$ bond angle. Are these bond angles the same?

6. What is the electronic shape around each carbon atom in ethylene?

7. How are single and double bonds different in VSEPR theory?

## VALENCE BOND THEORY

In VSEPR theory, counting the number of electron pairs around an atom allows you to predict bond angles and, thus, molecular shape. Although VSEPR theory is very good at predicting bond angles and electron pair geometry, it does not give an adequate explanation of double and triple bonds. Valence bond theory, on the other hand, provides a better description of the double and triple bonds that are often found in organic molecules.

8. Consider an ethane molecule $(C_2H_6)$ to answer the questions that follow.

   a) Consult your model for $C_2H_6$ that you built earlier. What types of hybrid orbitals are present at each carbon atom in ethane?

   b) Use valence bond theory to draw a diagram representing the bonding in ethane.

   c) How many carbon–carbon sigma bonds are present in ethane?

   d) Are the carbon–hydrogen bonds in ethane sigma bonds or pi bonds?

9. Consider a model of an ethylene molecule $(C_2H_4)$ to answer the questions that follow.

   a) What types of hybrid orbitals are present at each carbon atom in ethylene?

   b) Use valence bond theory to draw a diagram representing the bonding in ethylene. Label the sigma and pi bonds.

   c) How does the C–C bond length in ethane compare with the C=C bond length in ethylene?

   d) Write a statement that explains why ethylene is planar using VSEPR theory and a second statement explaining why ethylene is planar, using valence bond theory.

10. Plastic bags: Polyethylene $(-(H_2C-CH_2)_n-)$ is the polymer used in plastic bags. Sketch and describe the molecular geometry of the carbon atoms in polyethylene, and label any sigma bonds, pi bonds, or bond angles.

11. Paper bags: Paper bags are made of the polymer cellulose, which consists of the monomer glucose ($C_6H_{12}O_6$). The chemical structure of glucose is shown below, with the carbons numbered to help distinguish them from each other.

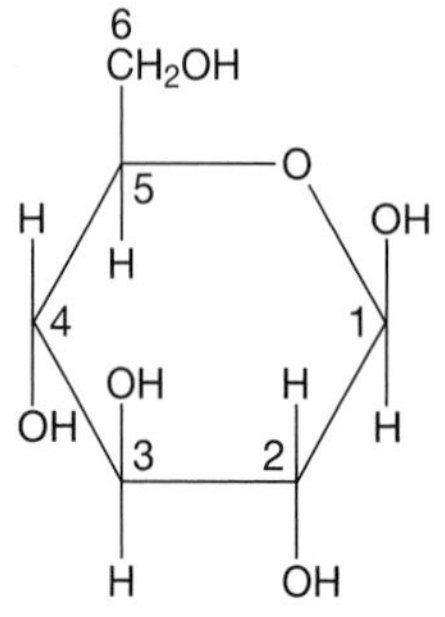

Glucose

a) What type(s) of hybrid orbitals are present at carbon number 2?

b) Would you expect the carbon and oxygen atoms in the ring all to lie in the same plane? Why or why not?

c) Build a model of glucose using a model kit. Does your model support your answer to part b? Explain.

12. On a separate sheet of paper, complete parts a through e for each of the molecules listed in the table on the following page. You may want to use molecular models to assist you.

a) Draw the Lewis structure for the molecule.

b) Name the molecular shape of the molecule and predict the "ideal" bond angles based on VSEPR theory. If the actual molecule would be expected to have bond angles that are different from these ideal angles, note what they are.

c) What types of hybrid orbitals would be found on the carbon atoms in each molecule?

d) Use valence bond theory to draw a diagram representing the bonding in the molecule.

e) How do the shapes and the bond angles predicted by valence bond theory compare with those predicted by VSEPR?

| NAME OF MOLECULE | FORMULA |
|---|---|
| Trifluorochloromethane | $CF_3Cl$ |
| Carbon dioxide | $CO_2$ |
| Ammonia | $NH_3$ |
| Hydrogen sulfide | $H_2S$ |
| 1,2 Dichloroethylene | $CHCl{=}CHCl$ |
| Benzene | |
| Adipoyl chloride (One of the monomers in nylon) | |
| Styrene (The monomer in polystyrene) | |
| Vinyl chloride (The monomer in polyvinylchloride) | $CH_2{=}CHCl$ |

Adapted by Sharon Anthony, Karen Harding, and Kim Kostka from the ChemConnections module "How Do We Get from Bonds to Bags, Bottles, and Backpacks?" by Karen Harding and Sharon Anthony.

# From Pesticides to Vitamins: Water Soluble or Fat Soluble?

- To practice different kinds of chemical notation, such as structural notation and line notation

- To practice identifying functional groups

- To apply learning about molecular polarity, intermolecular forces, and solubility to make predictions related to pesticides and vitamins

- To analyze real data to determine whether predictions are supported or not supported

## INTRODUCTION

The fact that "oil and water don't mix" is a well-known phenomenon. But why don't they mix? What features of chemical composition or structure determine how substances mix? We use the terms *miscible* and *soluble* to describe substances that mix with or dissolve each other, whereas *immiscible* and *insoluble* refer to substances that don't mix. But is solubility really such an all-or-nothing proposition? In this activity, you will explore how the polarity of a molecule can be used to explain and predict solubility.

## PART I: SOLUBILITY AND PESTICIDES

1. The condensed chemical structure for Alar (daminozide), a pesticide and plant growth regulator that was used on apples in the 1980s, is given here:

$$(CH_3)_2N\text{–}NH\text{–}CO\text{–}CH_2CH_2COOH$$

a) Draw the line notation for Alar.

b) Would you expect Alar to dissolve in water or fat? Explain your reasoning.

2. The following structure is dichlorodiphenyl trichloroethane (DDT), one of the first successful insecticides. After its introduction in the late 1950s, crop yields improved and the incidence of malaria reduced dramatically in countries where this disease had taken a high toll. DDT is readily absorbed by insects and is stable in the environment, so its effectiveness in fields and grain-storage areas is long-lived. However, this longevity is also a problem because DDT accumulates in the fatty tissues (liver, kidneys, skin) of animals and is passed up the food chain. Top predators such as hawks and eagles can accumulate high levels of DDT; some species have recovered only recently from the decline in reproductive success they suffered because of DDT's toxic effects. For these reasons, DDT has been banned in the United States and many other countries.

DDT

a) Based on its chemical structure, rationalize why DDT is long-lived in the environment and why it collects preferentially in the fatty tissues of animals.

b) Pesticides have more recently been designed to form polar products when metabolized by animals. For example, methoxyclor has a structure like that of DDT but with $OCH_3$ groups instead of Cl groups on the benzene rings, making it more digestible by bacteria in the gut. Explain why degrading into a polar product will help solve the problem of bioaccumulation of this pesticide.

3. Below are the chemical structures of two pesticides. Use your understanding of polarity and water solubility to determine which pesticide would be easier to wash off of produce. Explain your reasoning.

Malathion—an organophosphate insecticide

Toxaphene—a pesticide

## PART II: SOLUBILITY AND VITAMINS

In Part II, you will work in small groups to analyze ultraviolet (UV) spectra of vitamins in water and ligroin. Absorbance and wavelength data for the vitamins A, $B_1$, $B_2$, C, E, and K have been collected by Chia Goh and George Lisensky at Beloit College. The procedure they used is: A vitamin sample was placed into a cuvette containing 2 mL of water and 2 mL of ligroin (predominantly a mixture of saturated alkanes of $C_7$ through $C_{11}$), the layers were shaken, and the vitamin was partitioned between the solvents. The solution was then allowed to separate into two layers, and a spectrum was obtained for each layer.

### Preliminary Questions

4. The cuvette shown here contains 2 mL of water and 2 mL of ligroin. Which layer is which? (*Hint:* The density of water is 1.0 g/mL, whereas that of ligroin is about 0.65 g/mL.)

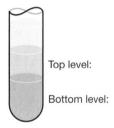

Top level:

Bottom level:

5. One way to classify vitamins is to specify their solubility in polar and nonpolar solvents. What do you know already about the solubility of the following vitamins? How soluble do you predict they will be in the two solvents used in this experiment? Make a prediction and record it in the table below. The common name for each vitamin is given along with its structure and chemical name.

| | | | Solubility in water | Solubility in ligroin |
|---|---|---|---|---|
| Vitamin A | CH₃ CH₃ H CH₃ H CH₃ —CH₂OH structure | Retinol | | |
| Vitamin B₁ | structure | Thiamine | | |
| Vitamin B₂ | structure | Riboflavin | | |

| | | | Solubility in water | Solubility in ligroin |
|---|---|---|---|---|
| Vitamin B$_6$ | | Pyridoxamine | | |
| Vitamin C | | Ascorbic acid | | |
| Vitamin D | | Vitamin D$_2$ (Calciferol) | | |
| Vitamin E | | α-Tocopherol | | |
| Vitamin K | | Vitamin K$_1$ (phylloquinone) | | |

6. Examine the UV spectra in the following figure or on your computer (colored versions of the vitamin spectra can be found at http://chemistry.beloit.edu/Fats/pages/vitamin.htm). The two plots in each figure show the absorbance of each vitamin in ligroin and the absorbance in water.

a) Which vitamins are more soluble in the ligroin?

b) Which are more soluble in the water?

c) Do these results fit with your predictions?

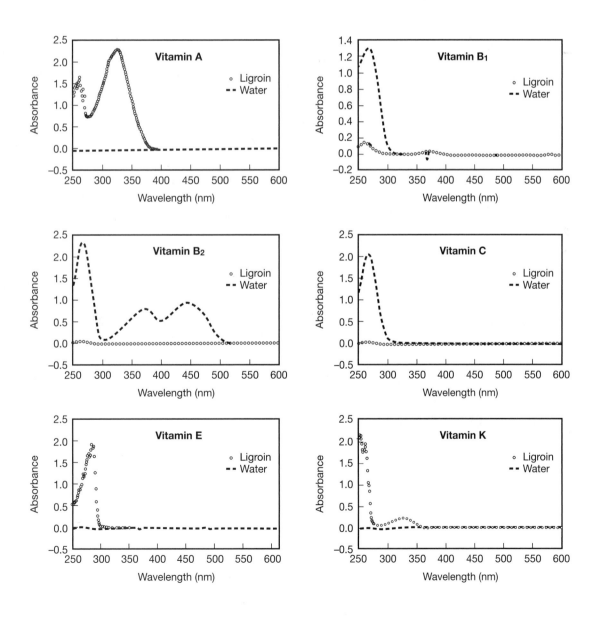

## Discussion Questions

7. Is it legitimate to call some vitamins water soluble and others fat soluble? Explain.

8. Many people believe that, because vitamins are good for you, more vitamins are even better! But taking megadoses of vitamins does not necessarily help you, and may even hurt. In fact, excess vitamin D can accumulate in and damage the kidneys; excess vitamin A accumulates in the liver and eventually damages it.

   a) In the body, vitamin D is concentrated in the skin, liver, and kidneys. Vitamin A is found in the liver and kidneys and in membranes such as the mucous membranes and the retina of the eye. Since excess vitamin A accumulates in the liver and excess vitamin D in the kidneys, what can you infer about the fat content of these tissues?

   b) Explain why excess amounts of vitamin C are excreted from the body in the urine, but excess amounts of vitamin A are not.

   c) Which vitamins are you most likely to need to consume every day, and why? Suggest why deficiencies of vitamins D, E, and K are fairly rare, but deficiencies of vitamin C and the B vitamins are much more common.

9. Interpret the solubility of vitamins A and C in water and ligroin in terms of the nature of the intermolecular forces between the molecules and the solvents.

## PART III: ADDITIONAL APPLICATIONS

10. Examine the structure of the molecule in this figure:

a) Would you expect this molecule to dissolve in water or fat? Explain your reasoning.

b) The preceding molecule is an example of a polychlorinated biphenyl (PCB). Use your understanding of how PCBs interact in the environment to check your answer to the previous question. (If you don't know much about how PCBs are found in the environment, do a little bit of research to check your answer.)

11. The structures of several alcohols are shown here. Rank them in order from most soluble to least soluble in water. Explain your reasoning.

*t*-Butanol    Ethanol    Isopropanol

Octanol

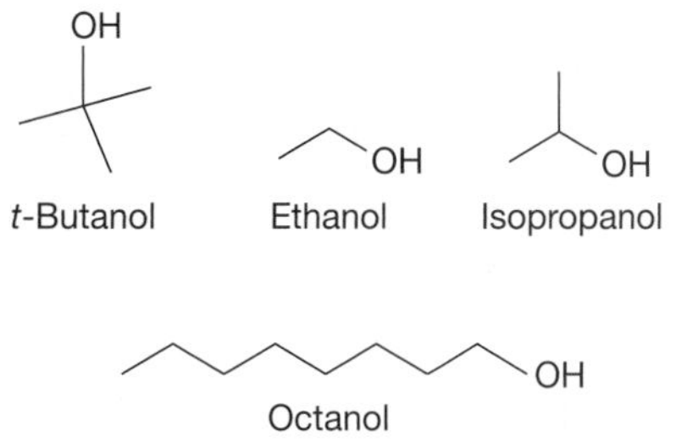

12. Capsaicin (shown here) is one of the ingredients that puts the heat in chili peppers. It stimulates nerve endings that detect pain in the mouth and elsewhere.

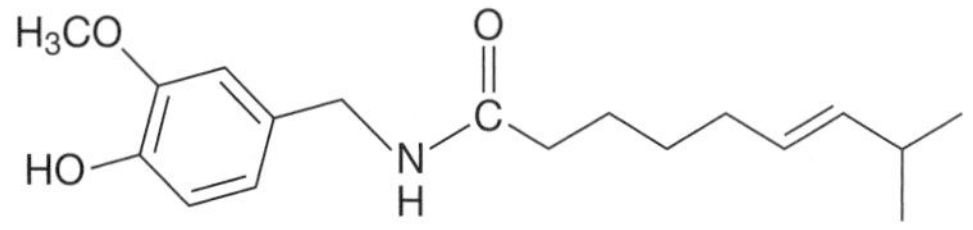

a) Predict whether capsaicin is more soluble in oil or in water. What consequences does this have for cooking with chili peppers?

b) Based on your answer to the previous question, which will help cool your mouth faster when you get a too-hot bite of Thai, Chinese, or Mexican food containing chili peppers: cold water or cold milk?

c) Capsaicin is also used as the active ingredient in muscle rubs that are used to treat arthritis or sore muscles. As a pharmaceutical chemist, would you plan to sell capsaicin in an oil-based cream or ointment, or in an isopropanol-based rubbing solution? Explain your reasoning.

d) Many flavor and aroma compounds are nonpolar molecules that dissolve preferentially in fat. How might this present a challenge in designing fat-free versions of popular foods?

Adapted by Heather Mernitz, Sharon Anthony, Kevin Braun, and Kim Schatz from the ChemConnections module "Would You Like Fries with That? The Fuss About Fats in Our Diet" by Sandra Laursen and Heather Mernitz.

# What Is the Difference between a Fat and an Oil?

- To analyze melting point data for a series of fatty acids and triglycerides

- To explain the melting point trends of fatty acids as a function of intermolecular forces and structure

- To categorize lipids based on their chemical structures: saturated versus unsaturated, monounsaturated versus polyunsaturated, cis versus trans

## INTRODUCTION

What is the difference between a fat and an oil? Fats have a higher content of saturated fatty acids, melt at higher temperatures, and are solid at room temperature. Fats tend to come from animal sources, although there are some plant-derived fats. Oils have more unsaturated fatty acids, melt at lower temperatures, and most are liquid at room temperature. Most oils come from plant sources.

In this activity, you will begin to explore how the chemical structure of fats and oils influences their melting point, the temperature at which the solid changes into a liquid. First, you will look at trends in melting points with respect to number of carbons and number of double bonds. Then you will build models of fatty acids to investigate how molecular structure influences melting point. The melting point of a triglyceride composed completely of a single fatty acid is very close to the melting point of the fatty acid itself. Thus, we can characterize the properties of the fat or oil by the properties of the fatty acids composing it.

## SELECTED CHEMICAL AND PHYSICAL PROPERTIES OF FATTY ACIDS

| Common Name | Condensed Structural Formula | Molecular Formula | Number of Carbon Atoms | Number of C–C Double Bonds | Fatty Acid Melting Point [°C] | Triglyceride Melting Point [°C] | Saturated, Monounsaturated, or Polyunsaturated? |
|---|---|---|---|---|---|---|---|
| Lauric acid | $CH_3(CH_2)_{10}COOH$ | $C_{12}H_{24}O_2$ | 12 | 0 | 44 | 46 | saturated |
| Myristic acid | $CH_3(CH_2)_{12}COOH$ | | | | 54 | 56 | |
| Palmitic acid | $CH_3(CH_2)_{14}COOH$ | | | | 63 | 65 | |
| Stearic acid | $CH_3(CH_2)_{16}COOH$ | | | | 70 | 71 | |
| Arachidic acid | $CH_3(CH_2)_{18}COOH$ | | | | 75 | 78 | |
| Behenic acid | $CH_3(CH_2)_{20}COOH$ | | | | 78 | 83 | |
| Lignoceric acid | $CH_3(CH_2)_{22}COOH$ | | | | 84 | 86 | |
| Myristoleic acid | $CH_3(CH_2)_3CH=CH(CH_2)_7COOH$ | $C_{14}H_{26}O_2$ | 14 | 1 | −4 | — | |
| Palmitoleic acid | $CH_3(CH_2)_5CH=CH(CH_2)_7COOH$ | | | | −0.1 | — | |
| Oleic acid | $CH_3(CH_2)_7CH=CH(CH_2)_7COOH$ | | | | 13 | −4 | |
| Gondoic acid | $CH_3(CH_2)_7CH=CH(CH_2)_9COOH$ | | | | 23 | — | |
| Linoleic acid | $CH_3(CH_2)_4(CH=CHCH_2)_2(CH_2)_6COOH$ | $C_{18}H_{32}O_2$ | | | −5 | very low | |
| α-linolenic | $CH_3CH_2(CH=CHCH_2)_3(CH_2)_6COOH$ | | | | −17 | — | |
| Arachidonic acid | $CH_3(CH_2)_4(CH=CHCH_2)_4(CH_2)_2COOH$ | | | | −49.5 | — | |

## DATA ANALYSIS: HOW DOES FATTY ACID MELTING POINT VARY WITH THE NUMBER OF DOUBLE BONDS AND NUMBER OF CARBON ATOMS?

1. Complete the preceding table, categorizing the fatty acids according to the number of carbons in the chain and the number of double bonds—two important features of the chemical structure. Some examples are filled in for you.

2. Now, look for trends in the data. Which fatty acids are solid at room temperature (about 25°C)? Which are liquid at room temperature?

3. How does increasing the number of carbons in the hydrocarbon chain affect the melting point? Find a set of fatty acids in the table—as large a set as possible—that shows the trend in melting point with the number of carbon atoms, *without being complicated by any other variables*. Make a graph to show this trend. What is the formula of the next fatty acid in the series? Use your graph to predict the melting point of this fatty acid.

4. How does increasing the number of double bonds in the hydrocarbon chain affect the melting point? Find a set of fatty acids in the table—as large a set as possible—that shows the trend in melting point with the number of carbon–carbon double bonds, *without being complicated by any other variables*. Make a graph to show this trend. What is the formula of the next fatty acid in the series? Use your graph to predict the melting point of this fatty acid.

5. Why is it important to group the data according to a single variable and to control the other variables?

6. According to these data, what factors influence the melting point of a triglyceride? Support your answer with examples.

## MODEL BUILDING: HOW CAN WE EXPLAIN THE RELATIONSHIP BETWEEN FATTY ACID MELTING POINT AND NUMBER OF DOUBLE BONDS?

7. Use a model kit to build the following structures:

   a) A 10-carbon saturated hydrocarbon

   b) A 10–carbon monounsaturated hydrocarbon with cis conformation

   c) A 10–carbon monounsaturated hydrocarbon with trans conformation

8. Draw the line notation for the molecules you built in Question 7.

9. Try twisting (without breaking bonds) the molecular models into different conformations. Can molecules rotate around single bonds? Can molecules rotate around double bonds?

10. Speculate whether saturated or monounsaturated fatty acid molecules would be able to pack closer together. Explain your reasoning.

11. How does the number of double bonds affect how closely molecules can pack together?

12. Predict how the closeness of molecular packing will affect the melting point of a substance. (*Hint:* What are the primary intermolecular forces between adjacent fatty acids and how will packing influence these forces?)

13. Summarize how the number of double bonds affects the melting point of fatty acids. Does this agree with the trends you observed in Question 2? Explain.

14. Examine your models of the cis and trans monounsaturated hydrocarbons. Using models, can you form one isomer from the other by twisting the molecules, or must you disconnect bonds?

15. Compare the molecular models of your cis and trans isomers. Which isomer do you think has the lower melting point? Explain.

16. How do you think the melting point of the trans hydrocarbon compares with that of its saturated counterpart? Explain.

## APPLICATIONS

17. Considering intermolecular forces, explain why it's not surprising that melting points for triglycerides with three identical R groups are about the same as the melting points of the fatty acids from which they are formed.

18. Epidemiological evidence shows that diets rich in fish such as cod, mackerel, and herring are associated with a low incidence of heart disease, even though such diets are not low in fat overall. These oily fish have a high content of a group of polyunsaturated fatty acids called *omega-3 fatty acids*, where *omega-3* indicates that one of the double bonds is three carbons away from the methyl end of the chain.

    a) Make a prediction about the melting point of polyunsaturated omega-3 fatty acids compared with that of saturated or monounsaturated fatty acids of the same chain length.

    b) Cod, mackerel, and herring live in cold water. Why is it advantageous for these fish to have a high level of omega-3 fatty acids in their cell membranes rather than saturated fatty acids?

Adapted by Heather Mernitz, Sharon Anthony, Kevin Braun, and Kim Schatz from the ChemConnections module "Would You Like Fries with That? The Fuss About Fats in Our Diet" by Sandra Laursen and Heather Mernitz.

# Laboratory: How Can We Make Oil and Water Mix?

- To understand how molecular polarity relates to the action of surfactants (soaps)

- To perform a saponification reaction (hydrolysis and acid–base reaction) in the laboratory and examine the product

## INTRODUCTION

The way that soap works to disperse grease is similar to the way that bile salts operate in the small intestine. The function of both these molecules depends on their having a polar end and a nonpolar end. The nonpolar end mixes with the fat globule, whereas the polar end projects into the watery medium, reducing surface tension and keeping the fat suspended in the water. Agitation breaks large fat globules into smaller ones, increasing the surface area exposed to digestive enzymes (or to your dishwater). Thus, these amphipathic molecules have special dual-solubility properties due to their unique combination of polar and nonpolar groupings.

The process for making soap using animal fat and a strong base derived from wood ash has been known since before the Roman Empire. Many American homesteaders also used this method. When lye (sodium hydroxide, NaOH) became commercially available, it was added to animal fat in a large kettle and cooked over a fire for several hours. The chemical reaction that takes place is called saponification, and it is the hydrolysis of the ester bonds in fats by a strong base. The products are sodium salts of the fatty acids—soap—and glycerol, an alcohol.

$$
\begin{array}{ccccc}
\overset{\displaystyle O}{\underset{\displaystyle \|}{}} & & & \overset{\displaystyle O}{\underset{\displaystyle \|}{}} \\
H_2C-O-C-R & & & H_2C-OH & NaO-C-R \\
\overset{\displaystyle O}{\underset{\displaystyle \|}{}} & & & & \overset{\displaystyle O}{\underset{\displaystyle \|}{}} \\
HC-O-C-R & + \quad 3\ NaOH & \longrightarrow & HC-OH & + \quad NaO-C-R \\
\overset{\displaystyle O}{\underset{\displaystyle \|}{}} & & & & \overset{\displaystyle O}{\underset{\displaystyle \|}{}} \\
H_2C-O-C-R & & & H_2C-OH & NaO-C-R
\end{array}
$$

A triglyceride     Strong base     Glycerol     Sodium salts of fatty acids

**153**

Because the resulting soap often contained excess unreacted NaOH, it would irritate the skin. Homemade soap thus has a reputation for being harsh, but use of excess fat instead of excess base—that is, careful attention to stoichiometry!—can make it much more pleasant to use.

## PART I: SOAP LABORATORY

### Safety Notes for All Procedures

*Warning*: NaOH is very corrosive. Avoid contact with eyes or skin and rinse any spills with plenty of water. Clean up any spills immediately. Heat is generated when NaOH is mixed with water; add the reactants slowly and watch for splattering. Make sure you are wearing safety goggles! Your instructor will provide details about which procedure, A or B, you will follow for making soap and alert you to any modifications. Both procedures use the same method for evaluating the properties of the final product.

### Procedure A: Making Soap for Personal Use

This procedure was adapted by Kim Kostka, University of Wisconsin–Rock County, from that outlined on a can of lye purchased at a hardware store. The final product is safe for use by most people, if the instructions are followed. If you wish to add materials to give the soap a particular fragrance or texture, bring them to lab with you. Only very small amounts are required.

#### Synthesis

The general procedure is given here. Use the table to identify the specific details of the procedure for the fat you are using. Coordinate with your classmates to try a number of different types of fats so that you can compare the properties of the soaps you made.

Accurately weigh 7–8 g (see the table) of sodium hydroxide pellets into a 150 mL beaker. Add 25 mL of deionized water and stir until the pellets are dissolved. The solution will become warm. Weigh out 50–60 g (see the table) of fat into a 250 mL beaker. Melt the fat by gently heating it, and adjust the temperature of the two solutions according to the table.

### Experimental Modifications for Making Soap from Different Fats

| Type of Fat | Amount of Fat (g) | Amount of NaOH (g) | Temperature of Fat Solution (°C) | Temperature of NaOH Solution (°C) |
|---|---|---|---|---|
| Lard (Pork Fat) | 55 | 7.5 | 29 | 24 |
| Tallow (Beef Fat) | 55 | 7.5 | 54 | 35 |
| Coconut Oil | 50 | 8.0 | 35 | 35 |
| Vegetable Oil or Shortening | 60 | 7.5 | 35 | 35 |

When the temperatures have been adjusted, pour the NaOH solution into the melted fat in a thin stream, with slow, even stirring. Too-rapid pouring or stirring causes separation. Stir slowly for 20–30 minutes, until the mixture becomes about the same thickness as honey.

At this point you may add small amounts of fragrance or essential oils—a drop or two is plenty!—and a pinch of dried botanicals (petals, ground spices, dried leaves, ground oatmeal, etc.). Take a small sample of the product to evaluate. Pour the mixture into a labeled paper or plastic cup or other suitable mold, cover it to retain the heat, and let it stand for 24 hours.

### Using Your Soap

Before you attempt to use the soap, remove it from the mold and let it cure by standing exposed to air in a dry place for at least two weeks, and preferably four weeks. Avoid using the soap on your face or other sensitive skin.

## Procedure B: Making Soap in Microscale Quantities

### Synthesis

Heat 100 mL of water in a 400 mL beaker on a hot plate until it begins to boil. While the beaker is heating, add 4.0 mL vegetable oil, 2.0 mL methanol, and 2.0 mL of 8-M sodium hydroxide to a screw-cap vial. Tightly screw on the lid to seal the vial. A leak, which will let reactant out or water in during the synthesis, will cause a small stream of bubbles to come from the vial cap in the water bath.

Place the vial in the hot water. After a minute or two of heating, the mixture should separate into layers, with the crude soap product formed at the interface between them. When this occurs, remove the vial from the hot water bath with a test tube holder and cool it slightly with tap water. Then wrap the vial in a paper towel to insulate your hands and shake it until a homogeneous solution is obtained, indicating that the reaction is complete. Return the vial to the hot water bath for 2–3 minutes to ensure that all the fat has reacted.

While the vial is heating, prepare an ice–salt bath, as follows. Add four heaping teaspoons of table salt (sodium chloride, NaCl) to 100 mL of cold tap water, add several ice cubes, and stir to dissolve the salt. Add more salt, if necessary, until no more salt dissolves. The addition of salt will make a mixture that is colder than an ice–water bath.

Remove the vial from the hot water and cool it slightly with tap water. If you see any solid, reheat the mixture to dissolve it completely. Pour the reaction mixture into the ice–salt bath and stir until the soap floats to the surface in a soap powder–like mass. Skim the precipitated soap from the surface, drain it, and collect it on a paper towel.

### Using Your Soap

Procedure B is optimized to be fast and simple rather than to produce a product that is sure to be mild when used on skin. It is not recommended that you wash your skin with this product.

---

This procedure was adapted from that of Walter Rohr, Eastchester High School, NY.

---

## PART II: EVALUATION OF SOAP PRODUCT

Record your observations and answers to the following questions in your laboratory notebook or in the space below.

1. Put some of the product in a test tube and shake it with a few milliliters of deionized water to check its solubility in water and its foam-forming ability. Add a drop of phenolphthalein. What does this probe tell you? NOTE: If the solution turns pink (i.e. is basic) the soap should not be used.

2. Put two drops of vegetable oil in a test tube with 10 mL of deionized water and observe. Now shake the tube and look for formation of an emulsion. Try the same thing with a small portion of added soap product. How does it compare with the first mixture? Can you see any evidence that your product is acting as an emulsifier? Compare the action of your soap in this test with that of a few drops of a solution of commercial soap or detergent.

3. How can you determine that the product is not just recovered starting material?

4. Soaps are soluble in water but fatty acids are not. Why? Add a bit of your product to deionized water and shake it until it is dissolved. Now add dilute hydrochloric acid dropwise. What do you observe? Write a chemical reaction to explain the formation of this product.

5. Look up the composition of the fat you used to make your soap. With this information, or by making reasonable assumptions based on what you already know about the molecular structure of fats, calculate whether the fat or the NaOH was in excess in this reaction.

6. Compare your observations with those of classmates who made soap from different fats. What similarities or differences in the properties of these products do you observe?

---

Adapted by Heather Mernitz, Kim Kostka, Kevin Braun, and Sharon Anthony from the ChemConnections module "Would You Like Fries with That? The Fuss About Fats in Our Diet" by Sandra Laursen and Heather Mernitz.

---

# How Can We Identify Metals?

## INTRODUCTION

The simple conductivity apparatus shown here consists of two wires that can be applied to a sample. If the sample is an electrical conductor, the circuit will be completed and the probe will light.

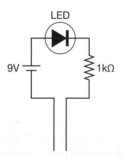

A schematic and photograph of a conductivity probe that lights when the circuit is completed by contact with an electrical conductor, such as a metal key.

A color version of the photograph can be found at www.wwnorton.com/college/chemistry/chemconnections/BlueLight/moviepages/Ag2.htm

**157**

# PART I: CLASSROOM ACTIVITY

1. Working singly or in groups, use a conductivity probe to find six things in the class-room that conduct electricity.

   a) Did anything enable you to quickly predict materials that would conduct electricity?

   b) Were any of your observations surprising? Did you find any nonmetals that did conduct electricity or any apparent metals that did not conduct electricity?

# PART II: WEB ACTIVITY

2. Use the link www.wwnorton.com/college/chemistry/chemconnections/BlueLight/pages/cond.html to observe a number of elements being tested for conductivity with a similar probe. As the conductivity or nonconductivity of an element is determined, mark conductors with one color and nonconductors with another color on the periodic table given here. The goal is *not* to check all the elements but to try to find the dividing line between metals and nonmetals in the periodic table.

| H | | | | | | | | | | | | | | | | | He |
|---|---|---|---|---|---|---|---|---|---|---|---|---|---|---|---|---|---|
| Li | Be | | | | | | | | | | | B | C | N | O | F | Ne |
| Na | Mg | | | | | | | | | | | Al | Si | P | S | Cl | Ar |
| K | Ca | Sc | Ti | V | Cr | Mn | Fe | Co | Ni | Cu | Zn | Ga | Ge | As | Se | Br | Kr |
| Rb | Sr | Y | Zr | Nb | Mo | Tc | Ru | Rh | Pd | Ag | Cd | In | Sn | Sb | Te | I | Xe |
| Cs | Ba | La | Hf | Ta | W | Re | Os | Ir | Pt | Au | Hg | Ti | Pb | Bi | Po | At | Rn |
| Fr | Ra | Ac | | | | | | | | | | | | | | | |

| | Ce | Pr | Nd | Pm | Sm | Eu | Gd | Tb | Dy | Ho | Er | Tm | Yb | Lu |
|---|---|---|---|---|---|---|---|---|---|---|---|---|---|---|
| | Th | Pa | U | Np | Pu | Am | Cm | Bk | Cf | Es | Fm | Md | No | Lr |

Mark the results of your conductivity tests on this periodic table. Which elements conduct electricity?

3. *Allotropes* are different structures of the same element that have different properties. Elements close to the division between metals and nonmetals may have both
conductive and nonconductive forms. Test the conductivity of carbon in its graphite
and diamond forms. Both are made entirely of the same element but differ in the
arrangement of atoms. Does the conductivity depend on the arrangement of atoms?

4. Conductivity also depends on impurities at the level of parts per million. For example, adding $H^+$ (from an acid such as HCl) or $OH^-$ (from a base such as NaOH)
to water increases the conductivity of the solution. Does the conductivity of silicon
samples vary with the levels of impurities?

# PART III: APPLICATIONS

5. Where are metal atoms found in the periodic table: on the left side or right side?

6. Predict whether each of these elements will conduct electricity:

   Potassium

   Rubidium

   Gallium

   Boron

   Bismuth

7. Based on your findings above, do any nonmetallic elements conduct electricity? If so,
name them.

Adapted by George Lisensky from the ChemConnections module "Build a Better CD Player: How Can You Get Blue Light from a
Solid?" by George C. Lisensky, Herbert Beall, Arthur B. Ellis, Dean J. Campbell, and Joanne Stewart.

# How Would You Design an Incandescent (Blackbody-Emitting) Bulb?

## LEARNING GOALS

- **To explore periodic properties of the elements**

- **To apply your knowledge of materials to design a lightbulb**

## INTRODUCTION

Materials must be heated to above 2000 K to produce a significant amount of blackbody visible light (see the following figure). The higher the temperature, the larger the fraction of power emitted as visible light and the closer the color match to the sun (approximately 6100 K blackbody emission). Not all materials can stand such a high temperature, though. The higher the temperature to which a lightbulb filament can be raised without failure, the better the filament. In this activity, you are asked to design an incandescent (blackbody-emitting) bulb that includes an appropriate filament that can be heated electrically.

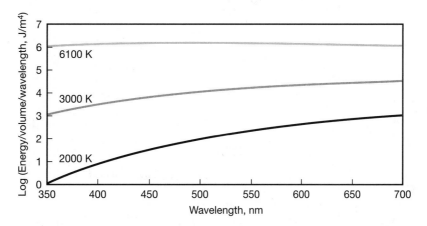

The visible portion of blackbody radiation, log curve.

As materials (including filaments) approach their melting point, the evaporation rate and the vapor pressure increase rapidly. (At higher temperatures, a larger fraction of molecules have sufficient energy to overcome the attractive forces of the solid and join the vapor

**161**

phase. Vapor pressure is a measure of the number of molecules in the vapor phase.) The evaporated material settles on the inside of the lightbulb, causing it to darken. The filament becomes thinner, increasing the resistance and thus increasing the temperature.

The filament must be uniform in diameter. Thin spots will be hotter because of their higher resistance, which leads to higher evaporation rates, further thinning, and eventual failure. (This is an example of positive feedback, in which an effect amplifies itself.)

In addition to melting, many materials oxidize (react with oxygen), and the rate of reaction increases with temperature. These materials burn when heated in the presence of oxygen. A few materials can be protected from further oxidation by a relatively impervious oxide layer. For example, the wire used in toasters, which is made of nichrome (80% nickel and 20% chromium), forms a layer of chromium oxide on its surface when the wire is heated in air. Oxygen diffusion through the chromium oxide is very low, even at high temperatures. Other elements with slow oxygen diffusion through the element's oxide coating include aluminum and rhodium.

Evaporation can be hindered by filling the bulb with an inert gas, but this also carries heat away from the filament and lowers the bulb's efficiency. The filling gas is often used at slightly less than atmospheric pressure, because the pressure will increase with temperature.

## PART I: PERIODIC PROPERTIES

Compare properties of the elements in the periodic table, and then use this information to choose elements for construction of an incandescent lightbulb.

1. Based on the following two figures, which of the following solid elements appear to be the worst candidates for lightbulb filaments? Why? In which portion of the periodic table are these materials found?

LOG (ELECTRICAL RESISTIVITY)

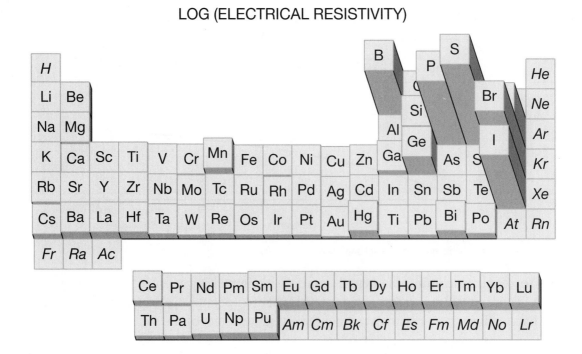

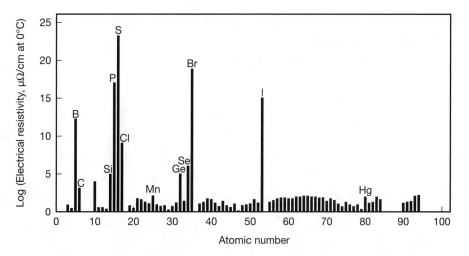

Gaseous elements cannot be made into wires, and their data are not shown. The taller the box or line, the larger the electrical resistivity.

2. Which of the following elements appear to be the best candidates for the connection between the filament and the wall socket? Why?

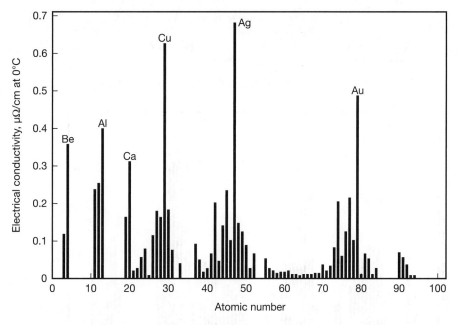

Gaseous elements cannot be made into wires, and their data are not shown. The taller the line, the better the electrical conductivity of the element.

3. The following figures compare the melting points of the elements in the periodic table.

a) Name the five elements with the highest melting points.

b) Are these elements also electrical conductors? The carbon data in the figure are for the graphite allotrope.

## MELTING POINTS

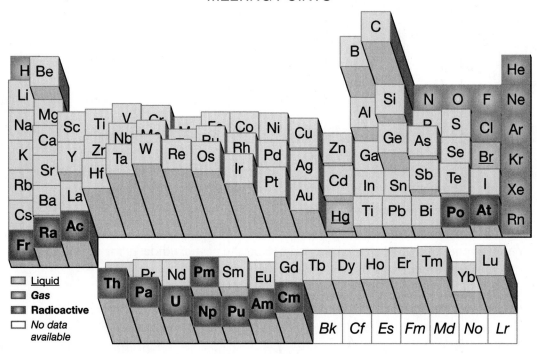

The taller the box, the higher the melting point. Gaseous and liquid elements have low melting points.

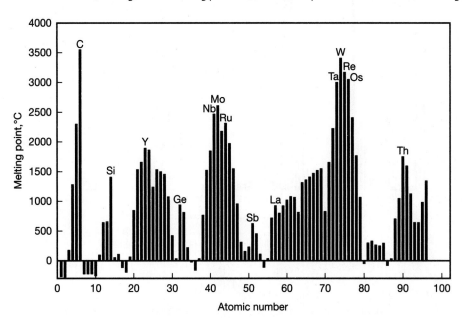

The taller the line, the higher the melting point. Gaseous and liquid elements have melting points below 25°C.

4. A semilog plot of vapor pressure as a function of inverse temperature for selected elements is shown in the next figure. Name two advantages of using a semilog plot for these data.

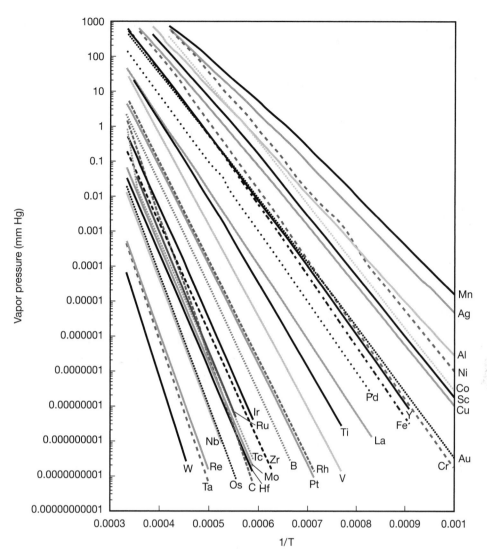

Vapor pressure in millimeters of mercury for some elements as a function of 1/temperature. The temperature scale covers the range from 3000 to 1000 K.

5. What would happen when current is applied to a lightbulb filament exposed to air? Watch the movie referenced in the figure and note your observations.

Oxidation of a lightbulb filament in air. The movie plays in slow motion.

This image is taken from a movie and can be viewed in color at www.wwnorton.com/college/chemistry/chemconnections/BlueLight/moviepages/tungsten.htm

# PART II: DESIGN AN INCANDESCENT (BLACKBODY-EMITTING) LIGHTBULB

6. What will be the operating temperature of your light? How did you choose?

7. Suggest three possibilities for the elemental composition of the filament. Why are these chemical elements good choices for a filament?

8. A semilog plot of vapor pressure as a function of temperature for selected elements is shown on the previous page. Estimate the vapor pressure of your filament choices at the operating temperature of your lightbulb.

9. What will be the size of the filament? Explain your reasoning. You can use relative terms such as "long" or "short" and "thick" or "thin" to describe size.

10. Suggest three possibilities for the composition of the connection between the filament and the electrical source. Why are these good choices?

11. What will be the size for the connection between the filament and the electrical source? Explain your reasoning.

12. Will a container be needed? If so, why? What properties are required for the container?

13. What temperature would be needed to get *blue* light from your incandescent bulb? Explain how you arrived at this answer. Would your filament survive?

## PART III: HALOGEN LAMPS

In a tungsten lightbulb, tungsten slowly evaporates from the hot filament and condenses on the glass or quartz window. *Halogen* bulbs contain a trace of a halogen such as iodine. Iodine vapor reacts with tungsten vapor:

$$W(g) + I_2(g) \rightleftharpoons WI_2(g)$$

$WI_2$ decomposes at the high temperature of the filament to release tungsten metal and iodine.

14. How does adding $I_2$ extend the life of an incandescent bulb at normal operating temperatures?

15. Why can halogen bulbs be operated at a higher temperature with the same lifetime as normal bulbs? (Operation at higher-than-normal temperatures also requires higher-melting quartz rather than glass for the window material.)

## References

Three-dimensional periodic table plotted by MacMendeleev Program, © 1987, Steven Sinofsky, Cornell University.

Resistivity, $\mu\Omega$-cm at 0°C, values from Dean, J. A. *Lange's Handbook of Chemistry*; McGraw Hill: New York, 1985.

Melting point, °C, values from Touloukian, Y. S.; Buyco, E. H. *TPRC Data series*; Plenum Press: New York, 1970; Vol 4.

Elemental vapor pressure data from Nesmeyanov, A. N. *Vapor Pressure of the Chemical Elements*; Elsevier: Amsterdam, 1963, pp 444–456.

Adapted by George Lisensky from the ChemConnections module "Build a Better CD Player: How Can You Get Blue Light from a Solid?" by George C. Lisensky, Herbert Beall, Arthur B. Ellis, Dean J. Campbell, and Joanne Stewart.

# How Does the Electrical Resistance of a Material Depend on Its Shape and Temperature?

## LEARNING GOALS

- To develop quantitative communication skills in graphing data

- To explore how the electrical resistance of a material depends on its shape and temperature

## INTRODUCTION

The electrical resistivity of a material inversely depends on the mobility of charge carriers, such as electrons and ions. It can vary over a large range, as shown by the logarithmic scale in the following figure.

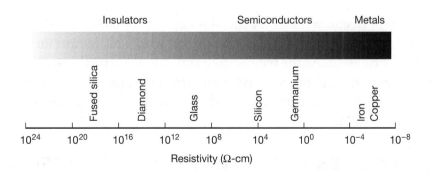

Multimeters commonly measure resistance, which depends on both the resistivity of the material and the shape of the material. Imagine water flowing through a pipe because of a difference in pressure between the ends of the pipe: Does it matter whether the pipe is long or short, narrow or wide, or empty or full of gravel? As we shall see, similar considerations apply for electrons flowing through a material because of a difference in voltage or electrical potential. We often classify materials as electrical conductors or nonconductors. Conductance equals 1 divided by resistance, so a low resistance means good conductance and vice versa. The unit of resistance is the ohm, represented by the Greek letter omega, $\Omega$. The unit of conductance is ohm$^{-1}$ or $\Omega^{-1}$.

**169**

Resistance is proportional to the length of the resistor, $l$, and inversely proportional to the resistor's cross-sectional area, $A$, which equals width times thickness. The proportionality constant, $\rho$, is called the *electrical resistivity*, which is a property of a material at a given temperature. Note that resist*ance* and resist*ivity* mean different things. The resistance, $R = \rho \dfrac{l}{A}$, is expressed in ohms. If $l$ is given in centimeters and $A$ in square centimeters, $\rho$ will have units of ohm-cm. The following table lists the electrical resistivities of some common substances. (The inverse of resistivity is called *conductivity*. The conductance of a material depends on its shape and its electrical conductivity.)

| Electrical Resistivities at 25°C for Some Metals | |
|---|---|
| Silver, Ag | $1.6 \times 10^{-6}$ ohm-cm |
| Copper, Cu | $1.7 \times 10^{-6}$ ohm-cm |
| Aluminum, Al | $2.8 \times 10^{-6}$ ohm-cm |
| Tungsten, W | $5.7 \times 10^{-6}$ ohm-cm [$92 \times 10^{-6}$ ohm-cm at 3000 K] |
| Zinc, Zn | $5.8 \times 10^{-6}$ ohm-cm |
| Iron, Fe | $10 \times 10^{-6}$ ohm-cm |
| Nichrome* | $100 \times 10^{-6}$ ohm-cm |

* Nichrome is an alloy of 80% nickel and 20% chromium.

The movement of electrons through a metal is impeded by collisions with the metal ions. The amplitude of the atomic thermal vibrations increases with temperature and increases the likelihood of a collision. As the temperature and atomic thermal vibrations increase, the resistivity increases slightly, as can be seen for tungsten in the table. The increase in resistance of a metal as the temperature increases is so characteristic of a metal that it is a diagnostic test for metallic behavior.

## PART I: HOW DOES THE RESISTANCE OF A MATERIAL DEPEND ON ITS DIMENSIONS?

In this activity, you will first measure the resistance of a graphite line as a function of its length, as in the top portion of the following figure. You will need a pure graphite #8B pencil (a normal #2 pencil has clay mixed in with the graphite and will not work as well), a piece of graph paper, and an ohmmeter or multimeter. You will obtain more reproducible results if you:

- Make your line reasonably dark.

- Keep track of units (the scale on the multimeter).

- Make all your measurements based on *one drawn line.*

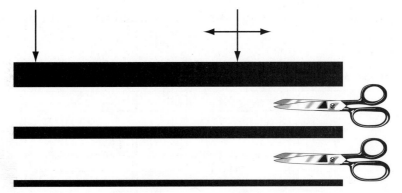

Steps in changing the line used for measuring resistance: [1] Start with a long wide line, [2] Measure its resistance as a function of length, [3] Cut that line to make progressively thinner lines, and [4] Measure their resistance as a function of length. *Do not draw more than one line!*

1. Check that your multimeter is measuring resistance (ohms, $\Omega$). When the leads are touching (you are measuring the resistance of the metal leads), the meter should read zero. When the leads are not touching (you are measuring the resistance of air), the meter should read infinity. What does your meter read when the leads are touching?

2. Draw a heavy line on the graph paper similar to the top line in the preceding figure. Use an alligator clip to connect one end of the graphite strip to the multimeter. Record the distance between leads and the measured resistance as you clip the other lead to the strip for at least five different lengths. (Do not cut the strip, just move the alligator clip along it.) Fill in the table, being sure to include units for each measurement. Note that the resistance scale may change!

**Width of strip:**

| Distance between Alligator Clips | Resistance |
|---|---|
|  |  |
|  |  |
|  |  |
|  |  |
|  |  |
|  |  |

3. Cut the graphite strip so that the width is thinner, and repeat your measurements.

### Width of strip:

| Distance between Alligator Clips | Resistance |
|---|---|
| | |
| | |
| | |
| | |
| | |
| | |

4. Once again, cut the graphite strip so that the width is thinner, and collect a third set of data.

### Width of strip:

| Distance between Alligator Clips | Resistance |
|---|---|
| | |
| | |
| | |
| | |
| | |
| | |

5. Plot resistance (ohms) versus length (cm) for each of your three trials on the same graph. What is the relationship between the length of a graphite line and the resistance of the line? (Examine your data or graph at a fixed width.)

6. What is the relationship between the width of a graphite line and the resistance of the line? (Examine your data or plot from Question 5 at a fixed length.)

7. Determine the average electrical resistivity of graphite using your measured values for resistance and the length and width of the line in centimeters. Use an approximate value of $2 \times 10^{-7}$ cm for the thickness of the line.

## PART II: HOW DOES THE RESISTIVITY OF A MATERIAL DEPEND ON ITS TEMPERATURE?

8. How much does the resistivity of 150 m of thin copper wire (a choke coil) change as it is cooled from room temperature to liquid nitrogen temperature ($-196°C$ or 77 K)?

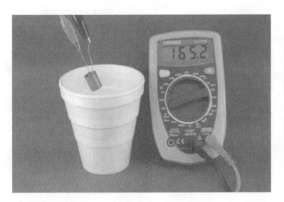 

Submersion of a choke coil in liquid nitrogen results in a change in resistance.

9. Place a 100 W lightbulb in a lamp. While the bulb is cool, measure the resistance of the lamp by connecting the meter leads across the power cord. What is the resistance?

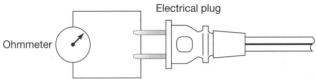

*Caution!* The lamp should *not* be plugged in when you make measurements.

If the resistance is very high, check that the on–off switch of the lamp is in the on position and measure again. Most of the measured resistance (~10 W) will be due to the bulb filament.

10. Why would it be desirable to have little resistance in the power cord?

11. Disconnect the meter leads. Plug the lamp into the wall socket and let the lightbulb heat to incandescence for several minutes. The temperature of the blackbody emitting filament will be approximately 3000 K. Remove the plug from the wall socket. *Immediately* measure the resistance (~150 W) of the lamp by connecting the meter leads across the power cord as before. In the following images, a double pole switch is used to disconnect the power and quickly connect the ohmmeter. Observe the resistance of the lightbulb filament as it rapidly cools. What do you notice?

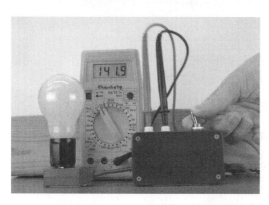

As a lightbulb filament heats up with use, the resistance of the lightbulb changes.

## PART III: APPLICATIONS

12. The 150 m length of copper wire in the choke coil used in Question 8 had a resistance of 150 Ω at room temperature. What is the diameter of copper wire?

13. A 120 V, 100 W lightbulb has a 0.0064 cm diameter tungsten filament and a resistance of 9.6 $\Omega$ at room temperature and 144 $\Omega$ at 3000 K. How long is the filament?

Examine the following figure to see how such long filaments can be made to fit in a normal lightbulb.

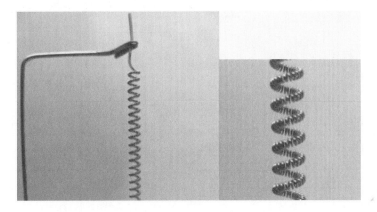

Close-up view of a tungsten 100 W lightbulb filament.

14. For a 120 V lightbulb, how do you make various power (wattage) lamps? (*Hint:* Combining $P = iV$ and $V = iR$ [Ohm's law] gives $P = V^2/R$.

15. Compare power consumption for the lightbulb in Part II when cold and when hot. When is a lightbulb most likely to burn out (when it is hot or cold)? Why?

16. How will stretching a wire (decreasing its diameter and increasing its length) change its resistivity? How will stretching a wire change its resistance?

17. Why does adding impurities to a metal increase its resistivity?

Adapted by George Lisensky from the ChemConnections module "Build a Better CD Player: How Can You Get Blue Light from a Solid?" by George C. Lisensky, Herbert Beall, Arthur B. Ellis, Dean J. Campbell, and Joanne Stewart.

# How Do Chemists Represent Large Molecules Found in Foods and Plastics?

- To understand and use several tools for representing molecules

- To interpret molecular structures for both small and large molecules

## INTRODUCTION

When you look at a chemical structure, you are observing a model that chemists use to describe the atoms in a molecule and how they are connected. Chemists have developed several shorthand tools to communicate this information. In this exercise, you will work with the structures of both large and small molecules and see that no matter how large or small, the tools remain the same.

## CHEMICAL NOTATION

### Molecular Formula

The molecular formula gives the number of atoms of each type in a molecule, such as $C_2H_6O$ for ethanol or $C_3H_6O_2$ for propanoic acid. It does not give any indication of how the atoms are arranged. For example:

| $C_2H_6O$ | $C_3H_6O_2$ | $C_{10}H_{22}O$ | $C_{10}H_{14}$ |
|-----------|-------------|-----------------|----------------|
| Ethanol | Propanoic acid | 1-Decanol | *t*-Butylbenzene |

### Structural Formula

The structural formula describes how the atoms are connected in the molecule. It is often used as a shorthand notation for Lewis structures. For example:

| Ethanol | Propanoic acid | 1-Decanol | *t*-Butylbenzene |

**177**

## Condensed Structural Notation

Condensed structural notation also describes how atoms are connected in a molecule, but it is simple enough to be written on one line. It is useful for efficiently describing the structure of large molecules with complex structures, such as polymers. The condensed structural notations for four compounds are shown here:

$CH_3CH_2OH$        $CH_3CH_2COOH$        $CH_3(CH_2)_9OH$        $C_6H_5C(CH_3)_3$

Ethanol                     Propanoic acid             1-Decanol                   *t*-Butylbenzene

## Line Notation

Line notation implies a carbon atom at each corner and at the end of a chain. Further, this notation assumes that a sufficient number of hydrogen atoms are present to satisfy the octet rule for each carbon atom, although these hydrogen atoms are not shown. All other elements and any hydrogen atoms associated with them must be shown. These can be drawn quickly and are able to convey a large amount of useful information. For example:

Ethanol    Propanoic acid                    1-Decanol                    *t*-Butylbenzene

## PRACTICE WITH NOTATION

1. In the following table, the molecular formula, the structural formula, and the condensed formula are given for 2-methyl pentane. Fill in the blanks for the rest of the table.

| Name | Molecular formula | Condensed structural notation | Structural formula | Line notation |
|------|------|------|------|------|
| 2-Methylpentane | $C_6H_{14}$ | $CH_3CHCH_3(CH_2)_2CH_3$ | | |
| 2,2-Dimethylbutane | | $CH_3C(CH_3)_2CH_2CH_3$ | | |
| 2,3-Dichloropentane | | | | |
| 1,3,5-Tribromohexane | | | | |
| 2-Chloro-2-methylbutane | | $CH_3CCICH_3CH_2CH_3$ | | |
| Pentanol | | | | |

2. Why might a chemist use a structural formula rather than a molecular formula?

3. Draw the structural formula and the line notation for each of the compounds listed.

a) $CH_3(CH_2)_5CH_3$

b) $C_3H_6$

c) $CH_3CH_2C(CH_3)_3$

d) $CClH_2CH_2Cl$

e) $CH_3CHCHCH_3$

4. Label the compounds from Question 3 as saturated or unsaturated hydrocarbons. Saturated hydrocarbons contain only carbon–carbon single bonds, whereas unsaturated hydrocarbons contain at least one carbon–carbon double bond or triple bond.

5. Many large biological molecules are found in nature. Fatty acids are large biomolecules that have the general structure shown here.

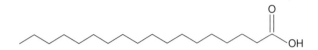

a) Draw the complete structural formula for this fatty acid, which is called *stearic acid.*

b) Draw the condensed structural formula for this fatty acid.

c) Is stearic acid a saturated fatty acid or an unsaturated fatty acid?

d) Draw a structure for any unsaturated fatty acid of your choice. (In nature, all fatty acids have an even number of carbon atoms.)

6. The polymer polyethylene, $-(H_2C-CH_2)_n-$, is made from many ethylene ($C_2H_4$) molecules that have bonded together.

a) Draw the structural formulas for ethylene and for a short section of polyethylene. Note that for polyethylene, because you are drawing only a small portion of a very large molecule, the octet rule will not be satisfied at the ends of the molecule.

b) Label each molecule as either a saturated or an unsaturated hydrocarbon.

c) Draw the line notations for ethylene and a short section of polyethylene.

Adapted by Sharon Anthony, Karen Harding, Kim Kostka, and Kim Schatz from the ChemConnections modules "How Do We Get from Bonds to Bags, Bottles, and Backpacks?" by Karen Harding and Sharon Anthony and "Would You Like Fries with That? The Fuss about Fats in Our Diet" by Sandra Laursen and Heather Mernitz.

ACTIVITY

37

# What Functional Groups Are in Foods?

## LEARNING GOALS

- **To gain practice recognizing and identifying functional groups in biomolecules**
- **To classify the three major macronutrients according to characteristic functional groups**

## INTRODUCTION

Nutrients can be classified according to their chemical structures. Their structures contain characteristic groups of atoms bonded in recognizable patterns, known as functional groups. In this activity, you will learn to recognize the three major classes of macronutrients—proteins, fats, and carbohydrates—by their characteristic functional groups.

## FUNCTIONAL GROUP IDENTIFICATION

Some functional groups common in food molecules are shown here. R is an organic group with one available bond. This is usually a carbon chain, but in some cases may be a hydrogen atom. You might think of it as the "rest" or the "remainder" of the molecule. R, R′, and R″ are often used to indicate that the groups may or may not be the same.

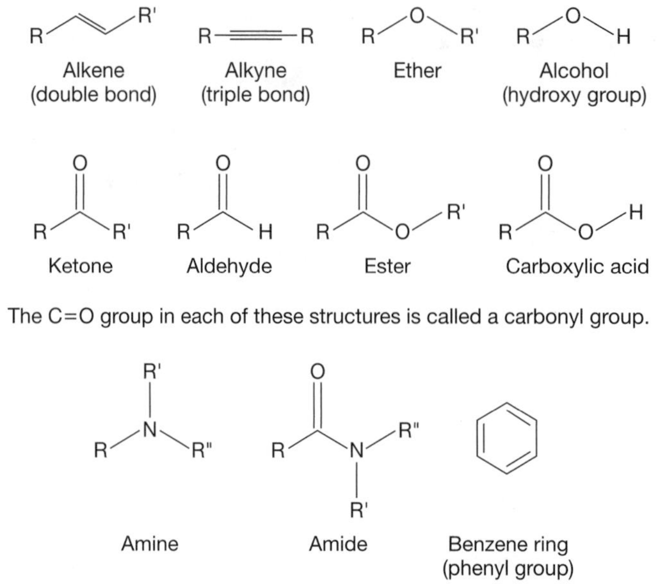

Alkene
(double bond)

Alkyne
(triple bond)

Ether

Alcohol
(hydroxy group)

Ketone

Aldehyde

Ester

Carboxylic acid

The C=O group in each of these structures is called a carbonyl group.

Amine

Amide

Benzene ring
(phenyl group)

Names and Structures of Functional Groups Common in Food Molecules

1. Study the structures of proteins, carbohydrates, and fats in the following figures. Using the preceding figure as your guide, find, circle, and label as many functional groups as you can.

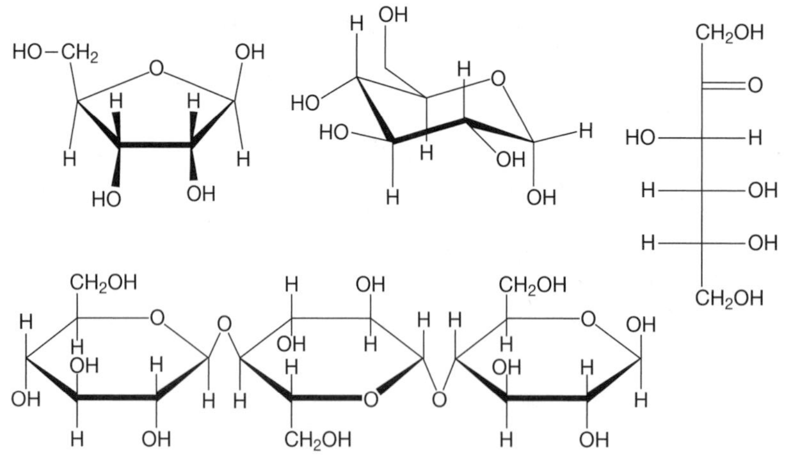

Carbohydrates

Proteins

Fats

2. What functional groups are typical of carbohydrates?

3. What functional groups do fats have in common?

4. What functional groups are common to proteins?

5. What structural features, besides the functional groups mentioned, seem to be typical of each class of macronutrient?

6. Identify the functional groups in each of the following molecules by circling and labeling them.

   a) Glucose

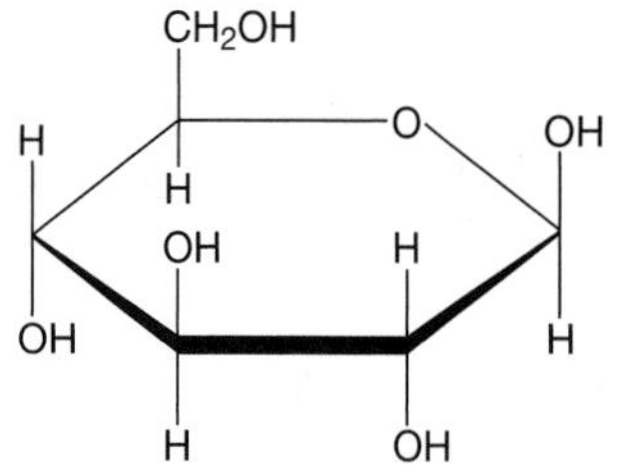

   b) Linoleic acid

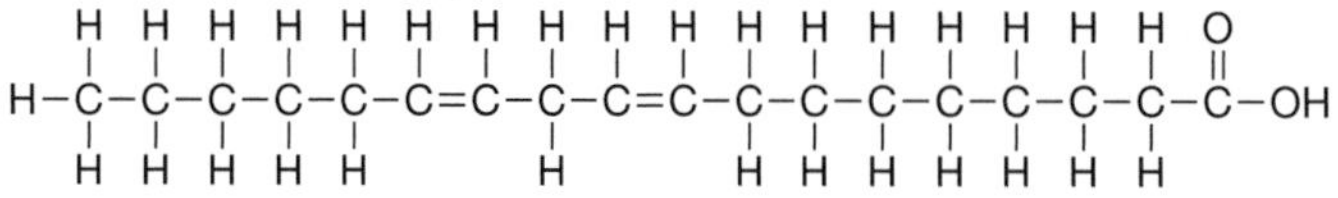

   c) Alanyl-serinyl-glutamic acid

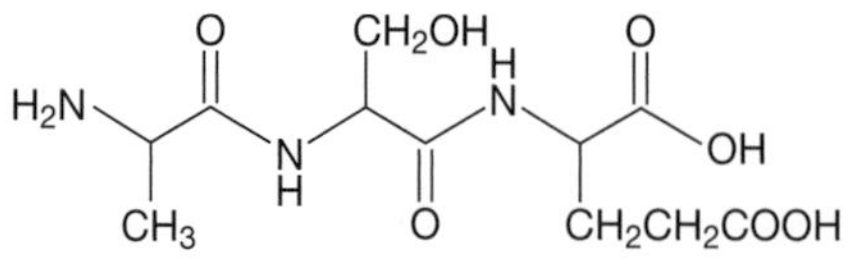

7. Assign each molecule shown in the previous figures to a macronutrient category (carbohydrate, protein, or fat) and list the structural features that help you identify each.

| Molecule | Macronutrient Category | Functional Groups and Structural Features |
|---|---|---|
| Glucose | | |
| Linoleic Acid | | |
| Alanyl-serinyl-glutamic Acid | | |

Adapted from the ChemConnections module "Would You Like Fries with That? The Fuss about Fats in Our Diet" by Sandra Laursen and Heather Mernitz.

ACTIVITY

# 38 How Are Fats Formed?

**LEARNING GOALS**

- **To practice using different types of molecular representation (model building, chemical notation) in building a triglyceride**
- **To explore the concepts of structural and geometric isomers**
- **To connect fat nomenclature to chemical structures: saturated versus unsaturated fats, monounsaturated versus polyunsaturated fats, trans fats**

## INTRODUCTION

All foods are broken into small molecules during digestion. Some of these small molecules are reassembled to form the more complex structures that the body needs, whereas others are metabolized—the biological equivalent of burning—to generate energy. Food molecules not immediately needed by the body are converted to triglycerides and stored in the adipose tissue for future energy production.

Fats, also known as triglycerides, are formed from the reaction of three fatty acids with a glycerol molecule, as shown here.

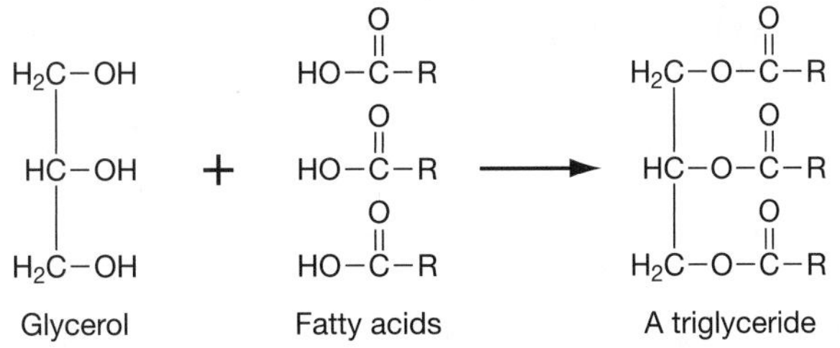


The partial reaction for the esterification of glycerol to form a triglyceride.

In triglycerides, the R group is a long carbon chain, 11 to 17 carbons long, with zero to three carbon–carbon double bonds. The R groups in a triglyceride can be all the same (although in nature this is rare) or different. Naturally occurring fats are typically made of only a few different fatty acids, some of which are listed in the table below. Compounds with only carbon–carbon single bonds in the hydrocarbon chain are called saturated because they have the maximum number of hydrogens attached. Carbon–carbon double bonds are referred to as sites of unsaturation because fewer than the maximum possible number of hydrogens are attached at these particular bonds. Fatty acids with one double bond are called monounsaturated, (the prefix mono means "one"), whereas fatty acids with more than one double bond are called polyunsaturated (the prefix poly means "many"). How does this technical meaning of the term saturated compare with the everyday meaning of the word?

# PART I: BUILDING A TRIGLYCERIDE

## Names and Formulas of Common Fatty Acids

| Category | Fatty Acid | Condensed Structural Formula | Molecular Formula |
|---|---|---|---|
| $C_{12}$ Fatty Acids | Lauric acid | $CH_3(CH_2)_{10}COOH$ | $C_{12}H_{24}O_2$ |
| $C_{14}$ Fatty Acids | Myristic acid | $CH_3(CH_2)_{12}COOH$ | $C_{14}H_{28}O_2$ |
| $C_{16}$ Fatty Acids | Palmitic acid | $CH_3(CH_2)_{14}COOH$ | $C_{16}H_{32}O_2$ |
| | Palmitoleic acid | $CH_3(CH_2)_5CH=CH(CH_2)_7COOH$ | $C_{16}H_{30}O_2$ |
| $C_{18}$ Fatty Acids | Stearic acid | $CH_3(CH_2)_{16}COOH$ | $C_{18}H_{36}O_2$ |
| | Oleic acid | $CH_3(CH_2)_7CH=CH(CH_2)_7COOH$ | $C_{18}H_{34}O_2$ |
| | Linoleic acid | $CH_3(CH_2)_4(CH=CHCH_2)_2(CH_2)_6COOH$ | $C_{18}H_{32}O_2$ |
| | Linolenic acid | $CH_3CH_2(CH=CHCH_2)_3(CH_2)_6COOH$ | $C_{18}H_{30}O_2$ |
| | Ricinoleic acid | $CH_3(CH_2)_5CH(OH)CH_2CH=CH(CH_2)_7COOH$ | $C_{18}H_{34}O_3$ |

1. Form a student group that will build a triglyceride using a molecular-model kit. Your group should first build three fatty acids (stearic acid, linoleic acid, and oleic acid) and a glycerol molecule (see the preceding table for fatty acid structures). Using your models, conduct a chemical reaction to form a triglyceride by building a triglyceride molecule from your three fatty acids and one glycerol molecule.

2. The chemical equation on the previous page is incomplete. What other product is formed in the reaction besides the triglyceride? Build it when you build your triglyceride.

3. How many different ways can you attach these same three fatty acids to the glycerol backbone? Compounds such as these with the same molecular formula but different structures are called **isomers**.

4. On a separate sheet of paper, draw the structure of the triglyceride model your group constructed.

5. Examine the previous table of fatty acids.

   a) Which fatty acids are saturated?

   b) Which are monounsaturated?

   c) Which are polyunsaturated?

## PART II: STRUCTURAL ISOMERS

As mentioned previously, compounds such as these, with the same molecular formula but different structures, are called isomers. There are structural isomers as well as geometric isomers, and each of these is important in studying fats. This activity will help you understand what a structural isomer is.

6. Draw a structural formula for a compound with the formula $C_4H_{10}$.

7. Using a model kit, construct a model that corresponds to your structural formula.

8. Rearrange the atoms in your model to make a different structure. (Do not eliminate any of the atoms, just rearrange them.) You will need to break some chemical bonds and form new chemical bonds to rearrange the atoms.

9. Draw the structural formula for this second molecule.

10. Your answer to Question 6 is a structural isomer of your answer to Question 9. Would you expect structural isomers to exist for a compound with the formula $C_5H_{12}$? If so, draw the structural formulas for each of the structural isomers. If not, explain why not.

## PART III: GEOMETRIC ISOMERS

Geometric isomers are also common with fats. You have heard of trans fatty acids. Now is your chance to learn what it means.

11. Use model kits to construct a model of 2-butene, $H_3CCH=CHCH_3$.

   a) How easily do the atoms rotate around a double bond compared with ease of rotation around a single bond?

   b) Without destroying your first model, build a different model for 2-butene. Each model should have a double bond between carbon 2 and carbon 3, but the models should not be identical. Draw the structural formula for each of your isomers.

   c) The two models you just built are called geometric isomers of each other. One of these is called the cis isomer and the other is called trans. Which is which? (Hint: In Latin, *cis* means "next to" and *trans* means "across.")

   d) Using models, can you form one isomer from the other by twisting the molecules, or must you disconnect bonds?

12. Find a fatty acid from the table in Part I that would form geometric isomers.

   a) On a separate sheet of paper, draw both the trans and the cis isomer in structural formula notation.

   b) On a separate sheet of paper, draw both the trans and the cis isomer in line notation.

Adapted by Sharon Anthony, Heather Mernitz, Kevin Braun, and Kim Schatz from the ChemConnections modules "Would You Like Fries with That? The Fuss about Fats in Our Diet" by Sandra Laursen and Heather Mernitz and "How Do We Get from Bonds to Bags, Bottles, and Backpacks?" by Karen Harding and Sharon Anthony.

Which functional groups do you see in this molecule? Circle and label each one you find.

12. This tripeptide is the result of the reaction of three different monomers, each of which is an amino acid. Which functional groups would you expect to find in an amino acid?

13. Write out the complete structure of the middle amino acid.

14. Using a reference tool or a biochemistry textbook, determine the name of the amino acid you drew in Question 13.

15. Would it be possible to make polymers out of amino acids? Why or why not?

16. Polypeptides are very important biological compounds, also known as *proteins*. How are polypeptides similar in structure to polyamides or polyesters?

Adapted by Sharon Anthony, Karen Harding, and Kim Kostka from the ChemConnections module "How Do We Get from Bonds to Bags, Bottles, and Backpacks?" by Karen Harding and Sharon Anthony.

# How Do Hot and Cold Packs Work?

## INTRODUCTION

Thermodynamics is a framework in chemistry that tells you whether a reaction will occur on its own. Reactions that, once started, will proceed on their own with no further input of energy are called spontaneous or exergonic. Reactions that need a continual input of energy to keep going are called nonspontaneous or endergonic. Two driving forces make chemical reactions happen: the formation of lower-energy products from higher-energy reactants in exothermic reactions ($\Delta H < 0$) and the formation of products that have more entropy than the reactants ($\Delta S > 0$).

You can predict whether a process will be spontaneous or nonspontaneous by relating the free energy available to do useful work (Gibbs free energy, $G$) to the enthalpy and entropy changes of the system.

$$\Delta G = \Delta H - T\Delta S$$

$\Delta G < 0$ for a spontaneous (exergonic) process.

$\Delta G > 0$ for a nonspontaneous (endergonic) process.

1. List three common reactions or processes that are spontaneous.

2. List three reactions or processes that are nonspontaneous.

Some types of hand warmers and warming pads use a slow combustion (oxidation) reaction. The same reaction occurs when rust forms in a drainpipe or on your bicycle.

$$4 \, Fe(s) + 3 \, O_2(g) \rightarrow 2 \, Fe_2O_3(s) + \text{heat} \qquad \Delta H = -1650 \text{ kJ}$$

3. If you open a packet of hand warmers and expose them to the oxygen in the air, you will feel the hand warmers becoming warm. Is this reaction spontaneous? How can you tell?

4. Given the value for $\Delta H$ for the reaction, is this an example of an exothermic or an endothermic reaction?

5. Why does the hot pack get hot? Explain in terms of heat flow between the system and the surroundings.

6. Examine the moles of reactants and products and the states of matter for each. Is entropy increasing or decreasing in this reaction? What is the sign of $\Delta S$?

7. Is this reaction favorable by enthalpy standards? Explain.

8. Is this reaction favorable by entropy standards? Explain.

9. Use the equation $\Delta G = \Delta H - T\Delta S$ to describe the spontaneity of this reaction. Make sure you compare the relative values of $\Delta H$ and $T\Delta S$.

10. Will this reaction be spontaneous at all temperatures? Explain.

Ammonium nitrate ($NH_4NO_3$), which is found in cold packs, is often used in fertilizers. In cold packs, it reacts with heat from the surroundings in the presence of water to dissociate into ammonium and nitrate ions.

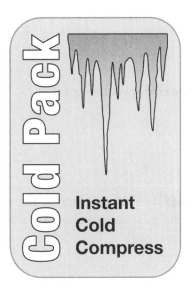

$$NH_4NO_3(s) + heat \xrightarrow{H_2O} NH_4^+(aq) + NO_3^-(aq) \qquad \Delta H = +28.1 \text{ kJ}$$

11. If you break open the water packet inside the cold pack and allow the ammonium nitrate and water to mix, you will feel the cold pack becoming cold. Is this reaction spontaneous? How can you tell?

12. Given that $\Delta H$ is +28.1 kJ for this reaction, is this an example of an exothermic or an endothermic reaction?

13. Why does the cold pack get cold? Explain your answer in terms of heat flow between the system and the surroundings.

14. Examine the moles of reactants and products and the states of matter for each. Is entropy increasing or decreasing in this reaction? What is the sign of $\Delta S$?

15. Is this reaction favorable by enthalpy standards?

16. Is this reaction favorable by entropy standards?

17. Use the equation $\Delta G = \Delta H - T\Delta S$ to describe the spontaneity of this reaction. Make sure you compared the relative values of $\Delta H$ and $T\Delta S$.

18. Will this reaction be spontaneous at all temperatures?

19. The idea of obstetric anesthesia for pain relief during childbirth gained acceptance after the administration of chloroform to Queen Victoria for the births of Prince Leopold and Princess Beatrice. Inhalation of chloroform vapors (gas) depresses the central nervous system. Chloroform liquid and gas are in equilibrium according to the following chemical equation:

$$CHCl_3(l) \rightleftharpoons CHCl_3(g) \qquad\qquad \Delta H = +6.98 \text{ kcal/mol}$$

a) Is this reaction exothermic or endothermic? Is heat a product or a reactant?

b) Does entropy increase or decrease in this reaction? What is the sign for $\Delta S$? How do you know?

c) Is the change of state from liquid to gas favored by $\Delta H$?

d) Is the change of state from liquid to gas favored by $\Delta S$?

e) Fill in the blank:
   The change of state from liquid chloroform to gas vapor is spontaneous _____.
   i.   at low temperatures
   ii.  at high temperatures
   iii. at all temperatures
   iv.  never

f) Explain your answer to Question e by discussing the sign of $\Delta G$ and its relationship to $\Delta H$ and $\Delta S$.

# How Do Automobile Emissions Contribute to Air Pollution?

## INTRODUCTION

Car exhaust plays a major role in the formation of smog, as found in Los Angeles and Houston. Two of the main reactants in the formation of smog are nitrogen oxides, such as NO, and carbon-containing compounds, such as CO and partially burned hydrocarbons. What thermodynamic forces drive the formation of these smog precursors? You will discover that these reactions proceed only under certain conditions, found in automobile engines. In this activity, you will explore what conditions are required to form these pollutants from molecules readily available in the atmosphere.

## PROBLEMS

| Standard Thermodynamic Values | | |
| --- | --- | --- |
| Formula | $S^0$ [J/mol·K] | $\Delta H^0$ [kJ/mol] |
| $N_2(g)$ | 191.5 | 0.0 |
| $O_2(g)$ | 205.0 | 0.0 |
| $CO(g)$ | 197.7 | −110.5 |
| $CO_2(g)$ | 213.8 | −393.5 |
| $NO(g)$ | 210.7 | 90.3 |

1. The smog precursor nitric oxide (NO) is formed in automobile engines via the following reaction:

$$N_2(g) + O_2(g) \rightleftharpoons 2\,NO(g)$$

a) Calculate $\Delta S°$ for this reaction at 25°C.

b) Calculate $\Delta H°$ for this reaction at 25°C.

c) Calculate whether this reaction would be spontaneous at room temperature, 298 K.

d) Calculate the equilibrium constant at room temperature, 298 K.

e) At what temperature does this reaction become spontaneous? Assume that neither $\Delta H°$ nor $\Delta S°$ changes substantially over this temperature range.

f) Calculate the equilibrium constant at this temperature.

g) Based on the equilibrium constants at 298 K and at the temperature you calculated in Question e, explain why NO forms in automobile engines but not in the atmosphere.

2. Another major contributor to smog formation, carbon monoxide (CO), is formed from the following reaction:

$$2\,CO_2(g) \rightleftharpoons 2\,CO(g) + O_2(g)$$

a) Calculate $\Delta S°$ for this reaction at 25°C.

b) Calculate $\Delta H°$ for this reaction at 25°C.

c) Calculate whether this reaction would be spontaneous at 25°C.

d) Calculate the equilibrium constant at room temperature, 298 K.

e) At what temperature does this reaction become spontaneous?

f) Calculate the equilibrium constant at the temperature from Question e.

3. Why are NO and CO not formed under ambient conditions, but only form under conditions found within an automobile engine?

Adapted from the ChemConnections module "How Can We Reduce Air Pollution from Automobiles?" by Howard Drossman, Wayne Tikkanen, and Sandra Laursen.

# Which Reactions Are Most Responsible for the Antarctic Ozone Hole?

- To understand why chlorine chemistry must be considered in understanding ozone depletion

- To write rate laws and do calculations using rate laws

## INTRODUCTION

Ozone ($O_3$), with its special ability to absorb ultraviolet (UV) radiation from the Sun, is often termed "Earth's sunscreen." Most of the "good" and protective ozone exists in the stratosphere at altitudes between about 10 and 30 km. In each Antarctic spring (September–October) since the mid-1980s, there has been a catastrophic loss of ozone in the Antarctic stratosphere above the South Pole centered around 15–20 km above sea level. At some altitudes, we observe the near complete absence of ozone. Massive ozone loss allows large amounts of damaging UV radiation to reach Earth's surface, leading to increased levels of skin cancer, genetic mutations, and other negative health effects on humans, as well as damage to the overall ecosystem. Scientists have gathered solid evidence to link the ozone hole to the human-made chlorine compounds called *chlorofluorocarbons* (CFCs). In Part I of this activity, you will analyze field data to see the relationship between ozone and chlorine in the Antarctic stratosphere.

Chemical rate studies, which produce kinetic data, are a primary tool scientists use to understand which reactions contribute most to ozone loss. Kinetic data have led the scientific community to a consensus understanding about causes of the ozone hole and how to fix it. In Part II, you will be working directly with kinetic rate data for several important reaction schemes to determine which chemistry is most responsible for the Antarctic ozone hole.

## PART I: WHY WORRY ABOUT CHLORINE?

Chlorine is often discussed in regard to ozone depletion, but why? To start, let's consider data obtained by an old spy plane that NASA converted to the ER-2 aircraft. Starting on September 16, 1987, from the tip of Chile (54°S), the aircraft began a multi-month mission to map the ozone hole.

The ER-2 flight data for September 16, 1987, follow. Note that the data labeled "Stratospheric chlorine" refers to chlorine monoxide, ClO, and that the scales differ on each side for the O$_3$ and ClO concentrations.

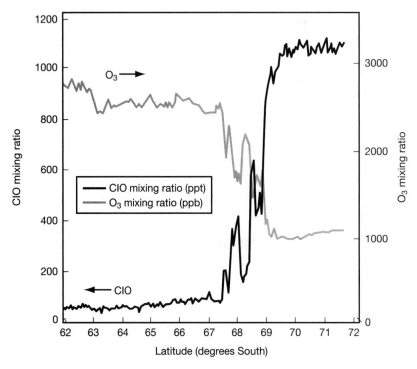

From Andersen, J. G.; Brune, W. H.; Proffit, M. H. Ozone Destruction of Chlorine Radicals within the Antarctic Vortex. *Journal of Geophysical Research* **1989**, *94*, pp 11465–11479.

1. Describe the trends in the ClO and O$_3$ data. Consider both general trends and the detailed structure in the data.

2. What are the concentrations for O$_3$ and ClO prior to and after the abrupt changes in the concentration data?

3. Carefully speculate about what the ER-2 September 1987 data set implies about Antarctic ozone destruction.

## PART II: WHICH CATALYTIC REACTIONS ARE MOST RESPONSIBLE FOR THE ANTARCTIC OZONE HOLE?

One natural oxygen reaction and two catalytic chlorine cycles play key roles in ozone destruction. The oxygen reaction (i) is part of the natural process of ozone destruction. The two-step chlorine cycle (ii) was proposed by Rowland and Molina in 1974 as a possible contributor. In the late 1980s, scientific field studies that detected ClO provided compelling evidence for both the two-step chlorine cycle and a second catalytic cycle, the ClO dimer cycle (iii). For both chlorine schemes, extensive field and laboratory studies now show that most of the chlorine originates from human-made CFCs.

| | | |
|---|---|---|
| i) Oxygen reaction | $O_3 + O$ | $\rightarrow 2O_2$ |
| ii) Two-step chlorine cycle | $Cl + O_3$ | $\rightarrow ClO + O_2$ |
| | $ClO + O$ | $\rightarrow Cl + O_2$ (rds) |
| iii) ClO dimer cycle | $2(Cl + O_3$ | $\rightarrow ClO + O_2)$ |
| | $ClO + ClO + M$ | $\rightarrow ClOOCl + M$ (rds) |
| | $ClOOCl + UV\ light$ | $\rightarrow ClOO + Cl$ |
| | $ClOO + M$ | $\rightarrow Cl + O_2 + M$ |

Note that $M$ is a collision partner that helps draw away evolving heat so products can survive. $M$ can be any atmospheric gas molecule and is most likely to be one of the more abundant species ($N_2$ or $O_2$).

In these reactions, the rate-determining step is abbreviated as *rds*. Under the cold sunlit conditions of the early Antarctic spring, ClO dimer formation is the rate-determining step (rds) for the dimer cycle (iii). Likewise, under the conditions of low oxygen concentration after a long dark winter in Antarctica, the second step in the two-step cycle (ii) is rate determining.

4. Write the *net* reaction for each of the reaction schemes listed above, cancelling out reagents that appear on both sides of the reaction as you would with a set of mathematical equations.

a)



b)



c)

5. For the three reaction schemes in Question 4, write the rate-law expressions for the destruction of ozone (with reaction orders and rate constants).

a)

b)

c)

6. For each of the three ozone-destruction schemes, calculate the rate constants and then the reaction rates for ozone loss at 190 K and 17 km using the Arrhenius equation and the following kinetic data and concentrations. The ozone hole is typically centered over Antarctica at an altitude of about 17 km, where the temperature is 190 K in early Antarctic spring.

**Concentrations**

$[ClO] = 2.4 \times 10^9$ molecules/cm$^3$

$[M] = 3 \times 10^{18}$ molecules/cm$^3$

$[O] = 4 \times 10^5$ molecules/cm$^3$

$[O_3] = 3.5 \times 10^{12}$ molecules/cm$^3$

| Gas-Phase Reaction | $k$ | $A$ $(cm^3 \cdot molecule^{-1} \cdot sec^{-1})$ | $E_a$ (kJ/mol) |
|---|---|---|---|
| $O_3 + O \rightarrow 2O_2$ | ——– | $8 \times 10^{-12}$ | $17.1 \pm 2.1$ |
| $ClO + O \rightarrow Cl + O_2$ | ——– | $3.0 \times 10^{-11}$ | $-0.6 \pm 0.6$ |
| $ClO + ClO + M \rightarrow ClOOCl + M$ | $7.2 \times 10^{-13}$* | ——– | ——– |

\* Second-order rate constant $T = 190$ K. The concentration of the third body, $[M]$, at 17 km and 190 K has already been included in this rate constant. Units are cm$^6 \cdot$molecule$^{-2} \cdot$s$^{-1}$.

a) Oxygen reaction
   i) Rate constant (with units)

   ii) Reaction rate (with units)

b) Two-step chlorine cycle
    i) Rate constant (with units)

    ii) Reaction rate (with units)

c) ClO dimer cycle
    i) Rate constant (with units)

    ii) Reaction rate (with units)

7. Which reaction scheme in Question 6 dominates the chemistry of the Antarctic stratosphere under cold sunlit conditions when catastrophic ozone depletion occurs? Explain your answer.

## References

Molina, M. J.; Rowland, F. S. Stratospheric Sink for Chlorofluoromethanes. *Nature* **1974**, *249*, p 810.

Andersen, J. G.; Brune, W. H.; Proffit, M. H. Ozone Destruction of Chlorine Radicals within the Antarctic Vortex. *Journal of Geophysical Research* **1989**, *94*, pp 11465–11479.

Portmann, R. National Oceanic and Atmospheric Administration Aeronomy Laboratory (private communication), March 1998. Concentrations based on Garcia–Solomon 2D model.

Data taken from Tables 1 and 2 in Evaluation Numbers 12 and 13, *Chemical Kinetics and Photochemical Data for Use in Stratospheric Modeling*, JPL Publications: Pasadena, Publication 97-4 **1997** (Jan 15), and Publication 00-3, **2000** (Mar 9).

Adapted by Sharon Anthony, Tricia Ferrett, and Kim Kostka from the ChemConnections module "Why Does the Ozone Hole Form" by Sharon Anthony and Tricia Ferrett.

# Why Is More Ozone Produced in the Stratosphere?

- **To gain experience with rate calculations**

- **To learn about the natural ozone cycle and how it gives rise to the ozone layer**

## INTRODUCTION

Earth's atmosphere is a complex and changing system. Though $N_2$ (78%) and $O_2$ (21%) are the most abundant molecules in the atmosphere, many other chemicals are present in trace amounts (less than 1%). Even though their atmospheric concentrations are low, trace species can still dramatically affect the chemistry of the atmosphere. However, just knowing the concentrations of trace gases is not enough. What matters is the extent to which chemical species actually participate in reactions that sustain or alter the chemical makeup of the atmosphere. To assess the impact of chemical species and their reactivity, it is necessary to know which chemical reactions dominate. The dominance of a reaction is governed primarily by how fast it occurs—fast chemistry is more dominant than slow chemistry, all else being equal. Analysis and calculation of the rate of a reaction is known as *kinetics*, a key tool used by many scientists. For example, the biochemist interested in how enzymes work in the body will usually measure reaction rates. Atmosphere chemists rely on studies of chemical reaction rates as a primary tool for uncovering how the atmosphere works.

One important trace chemical in the atmosphere is ozone $(O_3)$. In the lower atmosphere (troposphere, ~0–10 km in altitude), ozone is present as a pollutant and is toxic to humans. This "bad" ozone is often prevalent in cities where air pollution is generated from the transportation sector and other industrial activities. In the upper atmosphere (stratosphere, ~10–50 km), "good" ozone shields Earth by absorbing damaging ultraviolet (UV) solar radiation. The vast majority of atmospheric ozone resides in the stratosphere, centered at about 20 km, where it is produced naturally using molecular oxygen $(O_2)$ and oxygen atoms (O).

This activity asks you to calculate and then analyze ozone-production reaction rates to figure out why there is more ozone produced naturally in the higher-altitude stratosphere compared with the troposphere.

**213**

## PART I: WHY IS LESS OZONE PRODUCED IN THE TROPOSPHERE THAN IN THE STRATOSPHERE?

Consider the primary ozone-producing reaction in the atmosphere:

$$M + O_2 + O \rightarrow O_3 + M$$

$M$ is a necessary collision partner, typically a major atmospheric gas such as $N_2$ or $O_2$. $M$ helps to manage the energy evolved in this exothermic reaction, removing heat that is evolved so the $O_3$ molecule can survive as a product. Note that this reaction involves the collision of three bodies. In atmosphere chemistry, three-body or termolecular reactions are relatively common. For reactions in liquids, the solvent often plays the role of "heat sink."

1. Write the rate-law expression for the preceding ozone-production reaction, assuming the reaction is an elementary step at the molecular level.

2. Use the concentration data in the accompanying table to calculate the rates of ozone production based on the preceding reaction. Show all your work and report your final rates with units.

|  | Altitude 10 km (Troposphere) | Altitude 25 km (Stratosphere) |
|---|---|---|
| $P$, pressure (atm) | 0.265 | 0.025 |
| $T$, temperature (K) | 230 | 218 |
| $[M]$ (molecule/cm$^3$) | $8.5 \times 10^{18}$ | $8.3 \times 10^{17}$ |
| $[O_2]$ (molecule/cm$^3$) | $1.8 \times 10^{18}$ | $1.7 \times 10^{17}$ |
| $[O]$ (molecule/cm$^3$) | $1.2 \times 10^4$ | $8.3 \times 10^6$ |
| Rate constant, $k$, has units of cm$^6$ molecule$^{-2}$ s$^{-1}$ (from Reference 2) | $1.14 \times 10^{-33}$ | $1.28 \times 10^{-33}$ |

a) Rate of ozone production at 10 km (in the troposphere).

b) Rate of ozone production at 25 km (in the stratosphere).

3. Using your rate results from Question 2, compare the rates at 25 km (stratosphere) and 10 km (troposphere). Which rate is faster? What does this imply about the relative amount of ozone produced at each altitude, ignoring other chemistry?

4. We will look at how the concentrations of each of the reactants in the rate law changes to understand why the rate of ozone production is greater in the stratosphere than in the troposphere.

|  | M | $O_2$ | O |
|---|---|---|---|
| Does the concentration of reactant increase or decrease going from 10 km to 25 km? |  |  |  |
| How many times larger or smaller? (*Hint:* Consider concentration ratios.) |  |  |  |

5. Given that the rate constant ($k$) is about the same at both altitudes, which reactant ($O_2$, O, or M), is primarily responsible for the larger $O_3$ production rate? Justify your answer.

6. The major route for formation of oxygen atoms is by photochemical reaction of $O_2$ with sunlight:

$$O_2 + UV \text{ sunlight} \rightarrow 2\,O$$

Why do you think there is a lower concentration of oxygen atoms in the troposphere than in the stratosphere?

7. Briefly summarize your findings in several sentences. Why is more ozone produced naturally in the stratosphere than in the troposphere? In your summary, be sure to support your conclusions with concentration, rate, and altitude data.

## PART II: WHICH OXYGEN REACTION DOMINATES OZONE DESTRUCTION?

In the Antarctic spring, chlorine from chlorofluorocarbons is responsible for the vast majority of the ozone depletion in the "ozone hole." However, there is also *natural* ozone depletion involving only oxygen and ultraviolet radiation from the sun. These oxygen-only ozone destruction reactions include the following:

$$1) \quad O_3 + UV \rightarrow O + O_2 \qquad \text{sunlight}$$
$$2) \quad O_3 + O \rightarrow 2\,O_2 \qquad \text{with O atom}$$

In this activity, you will calculate which of these reactions dominates oxygen-only ozone destruction. The rates of these reactions are affected by sunlight, reactant concentrations, and rate constants, all of which vary with altitude. A careful rate analysis, using realistic and measured values at altitudes of interest, is needed to determine which reaction dominates natural oxygen-only ozone destruction.

8. For the oxygen atom–assisted reaction (2), write the rate law for the rate of destruction of ozone.

9. For reaction 2, calculate the rate of ozone destruction at 30 km (in the middle of the stratosphere) using the following data. Report units on your rate. The experimentally measured rate constant can be written in the Arrhenius equation form as:

$$k = 8 \times 10^{-12} \cdot e^{(-2060/T)}$$

with T in Kelvin. Units for $k$ are $cm^3$ molecule$^{-1}$ s$^{-1}$. You can see that the rate constant depends on temperature.

| | Altitude 30 km (Stratosphere) |
|---|---|
| $P$, pressure (atm) | 0.011 |
| $T$, temperature (K) | 224 |
| $[M]$ (molecule/cm$^3$) | $3.7 \times 10^{17}$ |
| $[O_2]$ (molecule/cm$^3$) | $7.8 \times 10^{16}$ |
| $[O]$ (molecule/cm$^3$) | $3.3 \times 10^7$ |
| $[O_3]$ (molecule/cm$^3$) | $3.1 \times 10^{12}$ |

10. The rate law for the UV-aided reaction (1) is written as

$$\text{rate of ozone destruction} = -\Delta[O_3]/\Delta t = J[O_3]$$

with $J = 6.5 \times 10^{-4} s^{-1}$ at 30 km. $J$ is a rate constant that varies with altitude, intensity of sunlight, angle of the Sun, and the inherent probability for ozone to absorb UV light. Thus, the use of $J$ leads to an approximate reaction rate. Use this $J$ information with concentration data from the table given in Question 9 to calculate the rate of ozone destruction from the UV-aided reaction at 30 km.

11. Compare the rates of ozone destruction you found for the oxygen atom–aided and UV-aided reactions. Which oxygen reaction dominates ozone destruction at 30 km and by what factor?

## References

1) Concentration data from the appendix in Brasseur, G.; Solomon, S. *Aeronomy of the Middle Stratosphere*, 2nd ed.; D. Reidel: Dordrect, Holland, 1986.

2) Rate constants derived from Table 2 in Evaluation Number 13, *Chemical Kinetics and Photochemical Data for Use in Stratospheric Modeling;* JPL Publications: Pasadena, Publication 00-3 **2000**, (Mar 9).

Adapted by Sharon Anthony, Tricia Ferrett, and Kim Kostka from the ChemConnections module "Why Does the Ozone Hole Form?" by Tricia Ferrett and Sharon Anthony.

# Why Is Chlorine an Efficient Ozone Destroyer?

■ **To apply the concepts of catalysis and reaction coordinate diagrams**

■ **To understand why chlorine chemistry is important in ozone depletion**

## INTRODUCTION

In this activity, you will examine a chlorine-catalyzed ozone-depletion reaction, and learn why chlorine is such an efficient ozone destroyer. You will also review some key chemical concepts in catalysis, including Arrhenius theory, activation energy, and the temperature dependence of rates.

## CHLORINE CATALYSIS

In 1974, Mario Molina and F. Sherwood Rowland proposed an important chemical mechanism for ozone destruction involving chlorine species. They suggested that an extensive catalytic chain reaction leading to the net destruction of $O_3$ and $O$ occurs in the stratosphere:

$$Cl + O_3 \quad \rightarrow \quad ClO + O_2$$
$$ClO + O \quad \rightarrow \quad Cl + O_2$$

This has important chemical consequences.

1. Write the *net* reaction for this two-step cycle, canceling out terms that appear on both sides of the reaction as you would with a mathematical equation.

2. Why don't Cl and ClO appear in the net reaction?

In the two-step mechanism in Question 1, Cl acts as a catalyst. A catalyst is a key player in the reaction; it is available to react at the beginning of the cycle and it is regenerated again at the end of the cycle. In essence, Cl is recycled so that it can destroy ozone over and over again.

In contrast, ClO is an intermediate. An intermediate is not present at the beginning or end of the cycle but is generated and consumed at an intermediate stage.

To understand catalysts, you need to look at the way in which they alter the energy pathway of a chemical reaction. As an example, compare the reaction energy diagram of both the uncatalyzed and chlorine-catalyzed reactions for ozone destruction.

3. First, analyze the uncatalyzed pathway for ozone destruction, as originally proposed by British physicist Sidney Chapman in 1930.

$$O_3 + O \rightarrow 2 O_2 \qquad \Delta H = -392 \text{ kJ/mol} \qquad E_a = 17.1 \text{ kJ/mol}$$

Draw the reaction coordinate diagram for the $O_3 + O \rightarrow 2 O_2$ reaction.

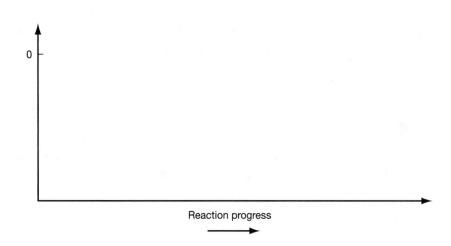

0

Reaction progress

4. Draw the reaction coordinate diagram for the chlorine-catalyzed pathway for ozone destruction. The $\Delta H$ and $E_a$ for each step are given here. (*Hint:* The diagram will have two humps.)

|  |  |  |  | $\Delta H$ (kJ/mol) | $E_a$ (kJ/mol) |
|---|---|---|---|---|---|
| Step 1: | $Cl + O_3$ | $\rightarrow$ | $ClO + O_2$ | $-155$ | $1.7$ |
| Step 2: | $ClO + O$ | $\rightarrow$ | $Cl + O_2$ | $-237$ | $\sim 0$ |
| Net: | $O_3 + O$ | $\overset{Cl}{\rightarrow}$ | $2 O_2$ |  |  |

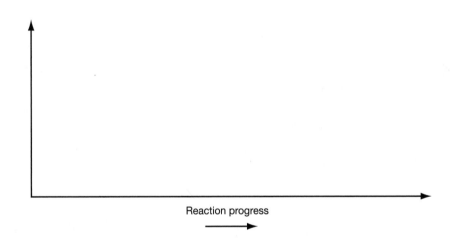

Reaction progress

5. Compare the activation energies for the catalyzed and uncatalyzed pathways. Which is larger, and is it twice as large? Five times larger? Ten times?

6. What do these reaction energy diagrams show about the role of a catalyst in reactions?

## Reference

Molina, M. J.; Rowland, F. S. Stratospheric Sink for Chlorofluoromethanes. *Nature* **1974**, *249*, p 810.

Adapted by Sharon Anthony and Kim Kostka from the ChemConnections module "Why Does the Ozone Hole Form?" by Tricia Ferrett and Sharon Anthony.

# What Happens to the Carbon Dioxide We Emit?

## LEARNING GOALS

- **To practice writing equilibrium constant expressions**
- **To apply Le Châtelier's principle**

## INTRODUCTION

You may have heard of carbon dioxide's role as a major contributor to climate change. Carbon dioxide also plays a major role in equilibria associated with rainwater, the oceans, and marine organisms. This activity introduces you to many of these important equilibria and asks you to determine how increasing concentrations of carbon dioxide will affect them.

## PART I: WHAT EQUILIBRIA DICTATE THE FORMATION AND SEQUESTERING OF CARBON DIOXIDE?

For each of the carbon dioxide source and sink reactions given here, write the equilibrium constant expression.

1. Respiration by animals and plants is a source of carbon dioxide. During respiration, a carbohydrate (e.g., glucose) reacts with oxygen to produce water and carbon dioxide.

$$C_6H_{12}O_6(s) + 6\,O_2(g) \rightarrow 6\,CO_2(g) + 6\,H_2O(l)$$

2. The combustion of fossil fuels is the main anthropogenic (human-made) source of carbon dioxide. In the United States, 44% of anthropogenic emissions are from petroleum, 35% from coal, and 19.5% from natural gas. During combustion, a hydrocarbon (e.g., octane, $C_8H_{18}$) reacts with oxygen to produce carbon dioxide and water.

$$2\,C_8H_{18}(l) + 25\,O_2(g) \rightarrow 16\,CO_2(g) + 18\,H_2O(l)$$

3. During photosynthesis, plants convert carbon dioxide and water to carbohydrates (e.g., glucose, $C_6H_{12}O_6$) and oxygen. Deforestation, particularly in tropical forests, contributes greatly to a reduction in photosynthesis as a carbon dioxide sink.

$$6\,CO_2(g) + 6\,H_2O(l) \rightarrow C_6H_{12}O_6(s) + 6\,O_2(g)$$

4. Carbon dioxide dissolves in water to yield carbonic acid. The oceans serve as a major sink for carbon dioxide. The transfer of carbon dioxide to the surface layers of the oceans occurs relatively quickly, on the order of a few years. However, it takes centuries for carbon dioxide from the atmosphere to reach the deep ocean.

$$CO_2(g) + H_2O(l) \rightleftharpoons H_2CO_3(aq)$$

## PART II: HOW DOES CARBON DIOXIDE AFFECT THE ENVIRONMENT?

Once it is released into the atmosphere, carbon dioxide readily dissolves in water, including rain, lakes, and the ocean. Once it is dissolved, carbon dioxide is involved in many secondary reactions that can lead to the sequestering of carbon dioxide or deleterious environmental effects.

5. The primary reaction describing the dissolution of $CO_2(g)$ in rainwater is

$$CO_2(g) + H_2O(l) \rightleftharpoons H^+(aq) + HCO_3^-(aq) + heat$$

   a) Write the equilibrium constant expression for this reaction.

   b) As the concentration of carbon dioxide in the atmosphere rises, how will this equilibrium shift?

   c) How will this affect the pH of rain?

6. In addition to rainwater, carbon dioxide also dissolves readily in the oceans according to the reaction presented in 5.

   a) If climate change causes the oceans to warm, what will happen to the ocean's ability to absorb carbon dioxide?

   b) About one third of the carbon dioxide emitted in the atmosphere is eventually absorbed by the oceans. Predict how rising carbon dioxide concentrations will affect the pH of the oceans.

   c) A major component of seashells is carbonate, which is created by the dissociation of bicarbonate formed when carbon dioxide dissolves in water. How will carbonate concentrations be affected by an increase in the acidity of the oceans?

   $$HCO_3^-(aq) \rightleftharpoons H^+(aq) + CO_3^{2-}(aq)$$

   d) Many marine organisms have protective shells composed of calcium carbonate. These shells are made according to the reaction:

   $$Ca^{2+}(aq) + CO_3^{2-}(aq) \rightleftharpoons CaCO_3(s)$$

   Write the equilibrium constant expression for the reaction and predict how the seashell formation reaction will shift as more carbon dioxide is emitted into the atmosphere.

7. Over time, geologic processes turn the calcium carbonate that accumulates on the sea floor into stone such as marble and limestone. These materials have been used as building materials for centuries. When exposed to acidic rain, stone that contains calcium carbonate weathers according to the reaction:

   $$CaCO_3(s) + 2 H^+(aq) \rightleftharpoons Ca^{2+}(aq) + H_2O(l) + CO_2(g)$$

   a) Write the equilibrium constant expression for this reaction.

   b) Explain how acid rain leads to the degradation of limestone buildings and statues.

# How Do Chemical Systems Respond to Stress?

## INTRODUCTION

Chemical equilibrium is a dynamic process in which reactants continually combine to give products, and products continually combine to give reactants. Eventually, a steady state situation is achieved, wherein the rate of production of products from the remaining reactants matches the rate of production of reactants from the remaining products.

Because equilibria are dynamic, if you change conditions such as concentrations, temperature, or pressure, the balance shifts. An increase in concentration shifts the balance toward the opposite side of the reaction. An increase in temperature alters the concentrations in the direction that absorbs heat. An increase in pressure alters the concentrations to favor the side with the smallest volume. The direction of the shift can be predicted by Le Châtelier's principle: A system at equilibrium responds to stress in a way that *partially* counteracts that stress.

## PART I: PREPARING AN EQUILIBRIUM SYSTEM

■ Place 30 red beans, 30 black beans, 30 black-eyed peas, and 30 white beans in a paper cup.

■ Reserve some additional beans to exchange when a successful reaction occurs.

■ Without looking, withdraw three beans from the reaction vessel. Exchange your beans according to the following reaction:

$$red + black \rightarrow 2\ white + black\text{-}eyed$$

**227**

■ Put the successful reactants aside; place products and unused reactants back in the reaction vessel. For example, if the three beans selected were two white and one black-eyed, you would exchange them for one red and one black bean. For example, if one red, one white, and one black bean were withdrawn, you would return a black-eyed and three white beans.

■ Keep a running tally of the number of each kind of bean in the cup. (Do this by adding or subtracting the changed beans rather than recounting all the beans in the cup.)

■ Record whether the reaction proceeded to the right or left ( $\rightarrow$ and $\leftarrow$ ).

■ If no reaction occurs, replace and try again. *Do not record unsuccessful reactions.*

■ Repeat until about half of your last 10 reactions go each way.

| Trial | Direction | Red Beans | Black Beans | White Beans | Black-Eyed Peas | K |
|-------|-----------|-----------|-------------|-------------|-----------------|---|
| 0 |  | 30 | 30 | 30 | 30 |  |
| 1 |  |  |  |  |  |  |
| 2 |  |  |  |  |  |  |
| 3 |  |  |  |  |  |  |
| 4 |  |  |  |  |  |  |
| 5 |  |  |  |  |  |  |
| 6 |  |  |  |  |  |  |
| 7 |  |  |  |  |  |  |
| 8 |  |  |  |  |  |  |
| 9 |  |  |  |  |  |  |
| 10 |  |  |  |  |  |  |
| 11 |  |  |  |  |  |  |
| 12 |  |  |  |  |  |  |
| 13 |  |  |  |  |  |  |
| 14 |  |  |  |  |  |  |
| 15 |  |  |  |  |  |  |
| 16 |  |  |  |  |  |  |
| 17 |  |  |  |  |  |  |
| 18 |  |  |  |  |  |  |
| 19 |  |  |  |  |  |  |
| 20 |  |  |  |  |  |  |
| 21 |  |  |  |  |  |  |
| 22 |  |  |  |  |  |  |
| 23 |  |  |  |  |  |  |
| 24 |  |  |  |  |  |  |
| 25 |  |  |  |  |  |  |

1. Has the color of the overall mixture changed? How?

2. Calculate the *K* value for your last few trials. Does your value for the equilibrium constant agree to the nearest power of 10 with that found by other class members? (To improve the statistics, you could start with a mole of each kind of beans and then run the reaction billions of times.)

## PART II: LE CHÂTELIER'S PRINCIPLE

Next you will investigate how this system responds to stress.

- Copy the last line of the previous table to the first line of the new table here.

- To stress the system, add an additional 10 more of one kind of bean to the reaction vessel and update your bean count.

- Withdraw three beans from the reaction vessel and react as before.

- Repeat until about half of your preceding ten reactions go each way.

| Trial | Direction | Red Beans | Black Beans | White Beans | Black-Eyed Peas | K |
|-------|-----------|-----------|-------------|-------------|-----------------|---|
| 0 | | 30 | 30 | 30 | 30 | |
| 1 | | | | | | |
| 2 | | | | | | |
| 3 | | | | | | |
| 4 | | | | | | |
| 5 | | | | | | |
| 6 | | | | | | |
| 7 | | | | | | |
| 8 | | | | | | |
| 9 | | | | | | |
| 10 | | | | | | |
| 11 | | | | | | |
| 12 | | | | | | |
| 13 | | | | | | |
| 14 | | | | | | |
| 15 | | | | | | |

*(continued)*

| Trial | Direction | Red Beans | Black Beans | White Beans | Black-Eyed Peas | K |
|-------|-----------|-----------|-------------|-------------|-----------------|---|
| 16 | | | | | | |
| 17 | | | | | | |
| 18 | | | | | | |
| 19 | | | | | | |
| 20 | | | | | | |
| 21 | | | | | | |
| 22 | | | | | | |
| 23 | | | | | | |
| 24 | | | | | | |
| 25 | | | | | | |

*Source:* Dickinson, P. D.; Erhardt, W. The "bean lab": A simple introduction to equilibrium. *Journal of Chemical Education* **1991**, *68*, p 930.

3. Did the initial reactions mainly go in the direction predicted by Le Châtelier's principle? Explain.

4. Calculate the *K* value for your last few trials. Is the same value of *K* obtained as before?

5. Compare your results with other class members who chose a different stress to apply to the system.

6. Explain your observations of the stressed bean systems in terms of Le Châtelier's principle. What stress was applied? How did concentrations change in response to that stress?

7. Explain your observations of the stressed bean systems in terms of the equilibrium constant. Was stress applied to the numerator or the denominator? How did the concentrations change to maintain the equilibrium expression as a constant?

Adapted by Sharon Anthony and George Lisensky from the ChemConnections module "Soil Equilibria: What Happens to Acid Rain?" by George Lisensky, Roxanne Hulet, Michael Beug, and Sharon Anthony.

# How Does Engine Temperature Affect CO and NO_x Production?

## LEARNING GOALS

- **To apply the concepts of Le Châtelier's principle to predict automobile emissions**

- **To understand how high engine temperatures and stoichiometry affect emissions of NO_x and CO**

## INTRODUCTION

Photochemical smog is formed when oxides of nitrogen react with organic carbon compounds in the presence of sunlight. Automobile engines are important contributors to photochemical smog because they release both carbon monoxide (CO) and oxides of nitrogen ($NO_x$). In this activity, you will examine how temperature affects the production of $NO_x$ and CO in automobile engines. The term $NO_x$ describes the oxides of nitrogen important in photochemical smog ($NO$ and $NO_2$). Because the equivalence ratio (a measure of the relative amounts of fuel and air) can be changed by the automobile exhaust-gas monitoring system to reduce pollution, there can be a compromise between the emission levels of different air pollutants. We will interpret the following figure, which was prepared by modeling the thermodynamic and kinetic behavior of an automobile engine as a function of the equivalence ratio.

1. In the following figure, an equivalence ratio of 1.0 represents the 1:1 stoichiometric ratio of fuel and air. Write the reaction for the combustion of isooctane ($C_8H_{18}$) in pure $O_2$. Which of the reactants and products are represented in the figure?

2. Dry air comprises 21% $O_2$ and 78% $N_2$ by volume (and molar ratio). Is the $O_2$ line at low equivalence ratios approximately 21% of all gaseous compounds in the engine? Draw a line representing the concentration of $N_2$ in the figure.

ChemConnections Activity Workbook

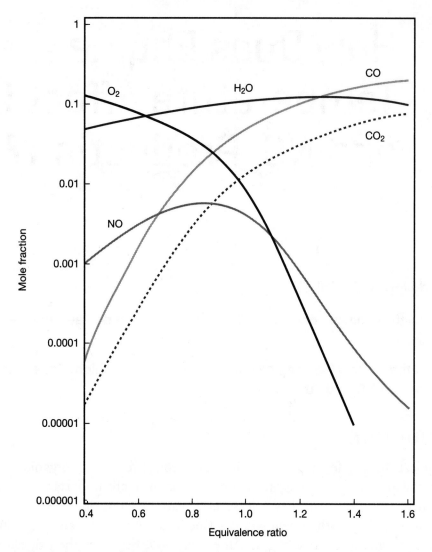

Equivalence ratio. The equivalence ratio indicates the relative ratio of fuel to air. The higher the equivalence ratio, the greater the ratio of fuel to air in the engine.

Figure adapted from data in Moor, A.R.; Heywood, J.B. Partial equilibrium model for predicting concentration of CO in combustion. *Acta Astronautica*, **1974**, *1*, p 952.

3. Based on the pattern of the O$_2$ mixing ratio in the figure, is there an excess of air or fuel at an equivalence ratio of less than 1.0? Explain.

4. A possible reaction of carbon dioxide is $2\,CO_2(g) \rightarrow 2\,CO(g) + O_2(g)$. Using your knowledge of stoichiometry and Le Châtelier's principle, hypothesize whether this reaction would be favored at higher or lower concentrations of $O_2$. Does this explain the pattern of the mixing ratio of CO in the figure? Explain.

5. Hypothesize whether the temperature of a combustion reaction would be highest at an equivalence ratio of 1.0 or some other equivalence ratio. Explain.

6. Explain why the highest NO mixing ratio is formed at an equivalence ratio of ~1.0.

7. If CO formation can be explained by Le Châtelier's principle using temperature *and* stoichiometric effects, hypothesize whether the $CO_2 \rightarrow CO$ reaction is endothermic or exothermic. Explain.

Adapted by Howard Drossman from the ChemConnections module "Can We Reduce Air Pollution from Automobiles?" by Sandra Laursen, Howard Drossman, and Wayne Tikkanen.

# Laboratory: How Can Used Fryer Oil Be Turned into Fuel?

- **To gain experience in organic synthesis by conducting a transesterification reaction**

- **To determine the efficiency in transforming triglycerides into methyl ester fatty acids**

- **To qualitatively determine the purity of synthesized biodiesel**

## INTRODUCTION

Over the preceding decade, the environmental impact of using fossil fuels, such as coal, petroleum, and natural gas for transportation and energy generation, is beginning to be fully realized. The nonrenewable nature of these energy sources, combined with the dramatic increase in greenhouse gases and pollution, has led scientists to explore alternative energy sources. An exciting alternative to fossil fuels is biodiesel synthesized from vegetable oils.

Unlike gasoline engines, diesel engines do not use spark plugs to ignite the fuel, but use high cylinder pressures to generate temperatures high enough to initiate fuel combustion. Diesel engines are generally more efficient and have better fuel economy than spark-ignited engines, though as with most combustion engines, they generate greenhouse gases, volatile organic compounds (VOCs), particulate matter, and carbon monoxide. The use of pure biodiesel or petrodiesel-blended biodiesel can substantially reduce these emissions. Advantageously, biodiesel can be used in modern diesel vehicles with little to no modification, though it should be noted that when burned, biodiesel exhaust can smell like french fries or popcorn.

Oils and fats are made of triglycerides, which are composed of three fatty acids connected to a glycerol molecule through an ester linkage (see the following figure). Most triglycerides contain two or three kinds of fatty acids of varying length and degree of unsaturation, which refers to the presence of double bonds in the fatty acid carbon chain. In oils and fats found within plants and animals, fatty acids typically contain 16 to 18 carbons, and over half the fatty acid residues are unsaturated (contain double bonds). Because a triglyceride

**237**

can contain three different fatty acids, the terms $R_1$, $R_2$, and $R_3$ are used to represent the carbon chains.

To synthesize biodiesel, the triglyceride molecules must undergo a *transesterification* reaction. As seen in the reaction scheme here, the ester groups attached to the glycerol backbone break and reform as three methyl esters in the presence of methanol ($CH_3OH$) and a base catalyst (KOH).

The base-catalyzed transesterification reaction of a triglyceride with methanol yields biodiesel and glycerol. Note that both the starting triglyceride and the produced biodiesel contain ester linkages.

## PART I: WASTE VEGETABLE OIL TITRATION

Because both fats and oils share the same basic chemical structure, biodiesel can be prepared from a range of starting materials. An attractive aspect of biodiesel is that it can be prepared from used oil, such as that found in your school cafeteria. To use waste vegetable oil (WVO), food particles must be filtered out and the acidity determined. When heated in the presence of water, a common environment for a cafeteria fryer, triglycerides decompose into free fatty acids and glycerol through a hydrolysis reaction. Because the transesterification reaction is base catalyzed, any free fatty acid can inhibit the reaction by neutralizing the catalyst. The transformation of your WVO into biodiesel will first require you to titrate an oil sample using a known concentration of sodium hydroxide (NaOH). The results of this titration will enable you to add enough base to both neutralize the WVO and catalyze the biodiesel formation reaction.

### Procedure

- Use a few milliliters of 0.01 M NaOH (note the precise concentration) to rinse your buret, then add about 10–15 mL NaOH and open the stopcock fully for a moment to clear any bubbles from the tip.

- To a small Erlenmeyer flask, add 10 mL of 2-propanol (isopropanol), 1.00 mL of WVO using an automatic pipet, and 3 drops of phenolphthalein indicator. Agitate the solution to mix.

- Note the buret's starting NaOH volume in your notebook and begin adding dropwise to the Erlenmeyer flask while swirling. The solution may turn cloudy; this is typical. Add NaOH until the solution turns light pink for at least 15 seconds. If a bright pink color is observed, you have gone past the endpoint and will need to start again. Note the final buret volume in your notebook.

- **Deposit all titration wastes in a labeled waste container.**

- Clean your buret with deionized water and leave it inverted with the stopcock open.
- Using the NaOH volume added and concentration, calculate the moles of fatty acids present in the 1.00 mL WVO sample. With the free fatty acid concentration, the moles of base required to neutralize the acid can be calculated.

## Calculations

1. The transesterification reaction requires 8.0 g of KOH catalyst per 1000 mL of pure vegetable oil. Calculate the moles of KOH required to catalyze 10 mL of WVO.

2. The KOH is provided as a 5.00 M solution in methanol. How many $\mu$L of 5.00 M KOH are required to catalyze 10 mL of WVO?

3. How many $\mu$L of 5.00 M KOH will be required to neutralize the free fatty acid in 10 mL of WVO?

4. Calculate the total volume of 5.00 M KOH to add to your 10 mL of WVO by summing the results of Questions 2 and 3.

## PART II: BIODIESEL SYNTHESIS

Now that you have calculated the volume of KOH required for the transesterification reaction, you are ready to make your own biodiesel.

## Procedure

- To a preweighed 25 mL Erlenmeyer flask add 10 mL of WVO using a syringe, and record the mass in your notebook. Add a small stir bar to the flask.

- In a small beaker, mix 2.0 mL of dry methanol and the volume of 5.00 M KOH calculated in Question 4. *Caution:* **Potassium hydroxide is caustic and methanol is flammable. Handle these substances with care.**

- Add the methanol–KOH solution to the 10 mL of WVO. Rapidly stir the solution at 55°C for 25–30 minutes on a stirring hotplate. The stirring rate should be high enough to thoroughly mix the components, but not too high as to cause the mixture to become frothy. Note any changes to the solution in your notebook.

- After the reaction is complete, pipet the solution into a 15 mL centrifuge tube. Either leave the tube upright until the next class period, or centrifuge the sample at 4000 rpm for 1 minute.

- Once separated, two distinct layers should be observed with glycerol and KOH catalyst in the bottom and biodiesel on top. Remove the bottom layer with a pipet. *Note:* **Deposit all waste in a labeled waste container.**

## PART III: BIODIESEL PURIFICATION

Before the synthesized biodiesel can be used as a fuel, it must be purified. Extraction steps must be performed to remove residual glycerol byproducts and KOH, flammable methanol, and fatty acid salts (*surfactants*) from your biodiesel. These residual chemicals and byproducts can adversely affect engine life and thus must be removed before use. Once the extraction steps are complete, all traces of water must be removed to prevent the formation of a catastrophic fuel–water emulsion.

### Procedure

- To your biodiesel, add 1 mL of 0.1 M acetic acid. Gently invert the tube (do not shake) for a few minutes, then allow the layers to separate (samples can be centrifuged to expedite the process). Discard the bottom layer.

- Repeat the process three more times using 1 mL aliquots of deionized water. Once again, discard the bottom layer. Note all observations in your notebook.

- Place the biodiesel in a preweighed 50 mL beaker and heat at 75°C–85°C for 20 minutes to remove any residual water. Alternatively, the biodiesel can be dried by adding a small spatula full of anhydrous calcium chloride ($CaCl_2$) or magnesium sulfate ($MgSO_4$) followed by filtration, using a syringe fitted with a filter.

- Once the biodiesel has cooled, measure the mass of the dry biodiesel and record it in your notebook. Because the starting WVO and biodiesel have roughly the same density, percent yield can be calculated from the approximation that 1 g of WVO should produce 1 g of biodiesel. What is the rough percent yield for your sample?

## PART IV: BIODIESEL ANALYSIS

Now that you have prepared your biodiesel, it is ready to be collected and poured into an engine, right? Before you do this, the biodiesel first must be tested to ensure its properties are similar to that of petroleum-based diesel and all of the triglycerides have been converted to methyl esters. Commercial standards for biodiesel are set by the American Society for Testing Materials (ASTM) International, whereas accreditation for commercial plants is provided through the National Biodiesel Accreditation Program called BQ-9000. Why is it important to have uniform standards for the biofuels industry?

You will perform a series of qualitative and quantitative tests to ensure the quality of your biodiesel. Record all of your observations in your laboratory notebook.

## Gel Test

The viscosity of biodiesel, like that of petroleum-based diesel, increases with decreasing temperature. If all the water has been removed, your biodiesel should gel when cooled. Why is it critical that the properties of your biodiesel emulate that of petroleum-based diesel?

- Transfer 2–3 mL of your biodiesel into a 15 mL centrifuge tube and place in an ice–salt bath or freezer for 15–20 minutes. Record the temperature of the biodiesel on removal. Stir the sample while holding it at the top of the centrifuge tube and record the temperature at which the sample clarifies.

- If your biodiesel is dry, it should gel during this period.

- If the biodiesel does not gel, reheat it at 75–85°C for 15–20 minutes or add an additional scoop of anhydrous $CaCl_2$ or $MgSO_4$ and filter it again.

- Once you are confident that your sample is dry, proceed to the next test.

## Percent Conversion Test (Methanol Test)

To verify the complete conversion of the triglycerides in the WVO to methyl esters, you will first perform the methanol test. This is the same test used by biodiesel homebrewers.

- To a 15 mL centrifuge tube, add 9 mL of dry methanol and 1 mL of purified biodiesel.

- Gently mix the tube by inverting it multiple times, then allow it to sit upright for 20 minutes.

- If you have produced a pure product, all of the biodiesel should dissolve in the methanol. Any unconverted WVO (triglycerides), though, will not dissolve and will quickly settle to the bottom. Record the volume of this layer, if formed, in your notebook.

- The volume of the unconverted layer can be used to calculate a rough percent conversion for your synthesis.

## Combustion Test

- Obtain two 2 dram vials, two 8 cm lengths of cotton string, and two 6 × 6-cm squares of aluminum foil.

- Add 1 mL of your biodiesel to one vial and 1 mL of WVO to the other. Submerge the string into the fuels using tweezers. Be sure the entire string is coated.

- Punch a small hole in the center of the foil, pull the string through it, and wrap the foil around the top of the vial. This will serve as the burner's wick. No more than 5 mm of the wick should extend above the top.

- In the fume hood, light the burners. Compare the burners for flame color and whether smoke or soot is produced. Was one burner easier to light? Is there any odor? Does one solution wick better? Measure how long it takes for the flames to self-extinguish.

## pH Test

- To 1 mL of deionized water, add 5 biodiesel drops and mix. Measure the pH of you biodiesel using a small strip of pH paper.

- Like petroleum-based diesel, your biodiesel should have a neutral pH of 7. Why is the pH of the biodiesel important?

## Structural Analysis: Fourier Transform Infrared Spectroscopy

Fourier transform infrared spectroscopy (FTIR) analysis provides a means to identify and quantify the composition of the final product.

- First, collect a background sample with nothing on the salt plate or attenuated total reflectance (ATR) window.

- Collect spectra of the WVO and your biodiesel.

- Overlay the two spectra plotted as absorbance versus wavenumber, label the prominent peaks, and print.

- Circle three prominent peaks in the combined spectrum. What functional group accounts for each peak?

- How does the biodiesel spectrum differ from that of the original fryer oil? Circle where the two differ. What structural change(s) would account for these differences?

## Advanced FTIR Analysis

*Note:* This test requires an ATR attachment.

To quantitatively determine the percent conversion (purity) of your biodiesel, you will compare the absorbance of your sample at 1435 cm$^{-1}$ (O–CH$_3$ stretch) against a series of standards of varying biodiesel concentration. The standards were prepared by mixing pure vegetable oil with 100% pure biodiesel by mass.

| Biodiesel Absorption at 1435 cm$^{-1}$ as a Function of Percent Conversion | |
| --- | --- |
| **Percent Conversion** | **Absorption at 1435 cm$^{-1}$** |
| 0 | 0.0317 |
| 20 | 0.0394 |
| 40 | 0.0469 |
| 60 | 0.0541 |
| 80 | 0.0622 |
| 100 | 0.0695 |

To determine the purity of your sample, plot absorption versus percent conversion. Using the equation of best fit, calculate the percent conversion of your sample using the absorption of your sample at 1435 cm$^{-1}$. Include the calibration plot in your paper with an X on the best-fit line at the absorption of your sample.

# PART V: REPORT

The purpose of your laboratory report is not only to describe your results but also to inform the reader of why this experiment was important and interesting. Be sure to address these key points in your laboratory report.

## Introduction

- Why was this experiment interesting and important?
- How did you synthesize biodiesel? (Include balanced chemical reactions.)
- How did you know whether you synthesized biodiesel?

## Experimental Methods

- Include detailed descriptions of the titration, synthesis, purification, and analysis steps.
- Be sure to note key observations.

## Results

- Include a table summarizing the titration results (moles of KOH required to neutralize the free fatty acids), moles of catalytic KOH, and total moles and volume of KOH added.
- Include a table reporting the results of the pH, gel, combustion, and percent conversion tests.
- Label and draw the functional groups associated with the key peaks in your FTIR spectra. Clearly denote which peaks change between the WVO and biodiesel.
- If applicable, show a graph with a best-fit line of absorbance at ~1435 $cm^{-1}$ versus percent biodiesel. The graph should include a point that represents the absorbance of your sample. Report the percent conversion in the figure caption.

## Discussion

- Did you produce biodiesel? If so, how do you know?
- What do the tests tell you about the quality of your biodiesel? In detail, explain what each test tells you about your biodiesel.
- If your percent conversion is below 100%, how might you improve the synthesis to obtain 100% conversion?
- If applicable: How does the percent conversion determined by FTIR compare with the methanol-based conversion test?

## References

WVO titration and gel test: Ed Brush and Fisher Science Education, *Green Chemistry in the Curriculum: Biodiesel Module.* www.fishersci.com/wps/downloads/segment/ScienceEducation/pdf/green_BiodieselModule.pdf

Synthesis conditions: http://make-biodiesel.org/

Combustion test: Bladt, D.; Murray, S.; Gitch, B.; Trout, H.; Liberko, C. Acid-Catalyzed Preparation of Biodiesel from Waste Vegetable Oil. *Journal of Chemical Education* **2011**, *88*, pp 201–203.

Advanced FTIR analysis: Mahamuni, N.; Adewuyi, Y. Fourier Transform Infrared Spectroscopy (FTIR) Method to Monitor Soy Biodiesel and Soybean Oil in Transesterification Reactions, Petrodiesel–Biodiesel Blends, and Blend Adulteration with Soy Oil. *Energy & Fuels* **2009**, *23*, pp 3773–3782.

Written by Kevin Braun and Brock Spencer

# What Is the pH of Household Chemicals?

- To calculate the pH of household chemicals

- To calculate the hydrogen ion and hydroxyl ion concentrations given the pH

## INTRODUCTION

Your household contains a wide variety of solutions including drinks, medicines, and cleaning products. In this activity, you will calculate the pH of some of these solutions. You will also calculate the hydrogen ion and hydroxyl ion concentrations of a solution with a given pH.

1. In orange juice, $[H^+]$ is approximately $1.4 \times 10^{-3}$ M.

   a) What is the pH?

   b) What is the $[OH^-]$?

   c) What is the pOH?

2. In typical household kitchen cleaners, the $[H^+]$ is approximately $3.2 \times 10^{-10}$ M.

   a) What is the pH?

   b) What is the $[OH^-]$?

   c) What is the pOH?

3. The pH of cola is approximately 2.5.

   a) Find the $[H^+]$.

   b) Find the $[OH^-]$.

4. Household ammonia has a pH of 11.5.

   a) Find the $[H^+]$.

b) Calculate the $[OH^-]$.

c) Calculate the pOH.

5. The hydroxyl ion concentration in red wine vinegar is $2.5 \times 10^{-12}$ M.

a) Calculate the $[H^+]$.

b) What is the pH?

6. A milk of magnesia solution has a hydrogen ion concentration of $4.4 \times 10^{-11}$ M.

a) Calculate the $[OH^-]$.

b) Calculate the pH.

7. The pH of wine varies greatly by type of wine, and it can also vary among batches. Vinny's Vineyard guidelines indicate that the pH of their wine must be between 2.9 and 3.4 for optimum taste. You are the quality control supervisor who determines whether a batch of wine leaves the vineyard. If a report on a sample comes back with a $[OH^-]$ of $3.5 \times 10^{-11}$ M, would you send the wine to market?

# What Chemicals Contribute to the Natural Acidity of Rainwater?

## LEARNING GOAL

■ **To apply equilibrium calculations to natural systems**

## INTRODUCTION

The acidity of rainwater is often blamed on anthropogenic sources such as nitric acid and sulfuric acid. It may come as a surprise, but there are other chemicals that significantly influence the acidity of rainwater. In this activity, you will use equilibrium calculations to explore the acidity of rainwater.

1. The primary reaction describing the dissolution of carbon dioxide in rainwater is:

$$CO_2(g) + H_2O(l) \rightleftharpoons H^+(aq) + HCO_3^-(aq) + \text{heat}$$

a) Write the equilibrium constant expression for this reaction. The equilibrium constant for this reaction, $K_a$, is $1.5 \times 10^{-8}$.

b) Notice that in the previous reaction, one $HCO_3^-$ is produced for every $H^+$. If this reaction is the primary source of $H^+$, then $[HCO_3^-]$ is equal to $[H^+]$. Substitute this information into the equilibrium constant expression from Part a so that $[HCO_3^-]$ no longer appears in your equilibrium constant expression.

c) The graph below shows the partial pressure in atmospheres for carbon dioxide gas as a function of the year. Read the most current value for $[CO_2(g)]$ from the graph and substitute it into your equilibrium constant expression to yield an equation with $[H^+]$ as the only unknown.

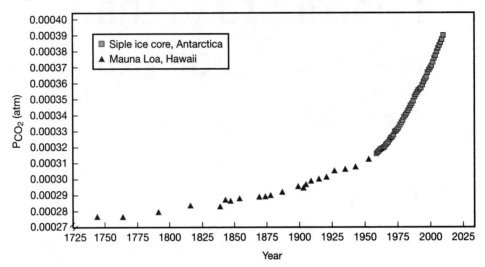

d) Using the equilibrium expression from Part c, solve for the current $[H^+]$ and pH of normal rain at equilibrium with the atmosphere.

e) Calculate the $[H^+]$ and pH of normal rain at equilibrium with the atmosphere in the year 1750.

f) Why do the $[H^+]$ concentrations in normal rain in 1750 and the present differ, assuming that no acid rain is present?

g) Assume that the concentration of $CO_2$ doubles in the next 50 years. Calculate the new pH of rainwater.

2. Formic acid and acetic acid are commonly found in rainwater. However, when pH monitoring stations analyze weekly bucket samples of rainwater, they do not detect these organic acids because they are decomposed by bacteria before the samples can be analyzed.

a) Write the chemical equation for formic acid, HCOOH, dissociating in water.

b) The typical concentration of formic acid in rainwater is approximately $3.7 \times 10^{-6}$ M. Calculate the pH of rain droplets, assuming that formic acid is the only acid present. The $K_a$ for formic acid is $1.8 \times 10^{-4}$.

c) As the formic acid is decomposed by bacteria, what happens to the $[H^+]$ in rain droplets?

d) Acetic acid, $CH_3COOH$, has a $K_a$ of $1.8 \times 10^{-5}$ M and an estimated concentration in rainwater of $1.4 \times 10^{-6}$ M. Write the formula for the dissociation of acetic acid in water.

e) Calculate the pH of rainwater, assuming that acetic acid is the only acid present.

f) Does formic acid or acetic acid contribute more to the acidity of natural rainwater?

# What Are the Buffer Systems in Your Body that Keep You Alive?

- To explore the definition and purpose of buffers

- To relate pH to $pK_a$ (Henderson–Hasselbalch equation) and investigate the role of buffers in the human body

## INTRODUCTION

What would happen if the pH of your blood or of the inside of your cells were to drop dramatically? One probable consequence is that many of the enzyme-catalyzed reactions in your body would slow significantly. Most enzymes in our bodies function only under very precise conditions. If blood or cellular pH strays too far outside of its normal range, enzyme activity decreases. Eventually, enzymes can become denatured, permanently disabling their activity. This situation can lead to illness and even death. Large changes in pH are prevented in the body by buffers, solutions that have the capacity to resist changes in pH by neutralizing small amounts of acid or base. In this activity, you will explore the three major buffer systems in your body that maintain physiological pH within a very narrow range.

Consider a beaker of pure water (pH = 7.00) to which either a strong acid (Beaker A) or a strong base (Beaker B) is added.

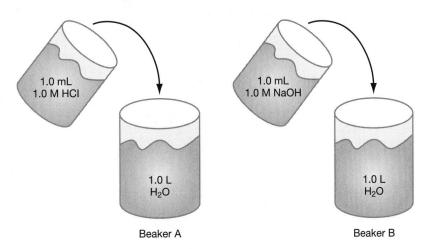

Beaker A                                    Beaker B

1. Calculate the pH in Beaker A.

2. What happens to the pH when a strong acid is added to the water?

3. Calculate the pH in Beaker B.

4. What happens to the pH when a strong base is added to the water?

Consider the following equation, showing the reaction between dihydrogen phosphate and water:

$$H_2PO_4^-(aq) + H_2O(l) \rightleftharpoons HPO_4^{2-}(aq) + H_3O^+(aq)$$

5. Label the weak acid and the conjugate base in the acid dissociation reaction.

6. Write the acid dissociation constant, $K_a$, for the dissociation of dihydrogen phosphate in water.

17. Le Châtelier's principle can be used to explain the action of buffer systems. Review the relevant sections of your textbook and your class notes, and write an explanation of Le Châtelier's principle here in your own words.

18. Certain conditions, such as asthma, pneumonia, emphysema, and smoke inhalation, lead to decreases in breathing rate. This results in an increase in dissolved $CO_2$ in the blood. Using the equation below and Le Châtelier's principle, explain whether this would result in a decrease or an increase in blood pH (clinically referred to as *acidosis* or *alkalosis*, respectively).

$$H_2O(l) + CO_2(g) \rightleftharpoons H_2CO_3(aq) \rightleftharpoons HCO_3^-(aq) + H^+(aq)$$

19. Metabolic acidosis is the decrease in blood pH that occurs when excessive amounts of acidic substances are not removed from the blood. This can occur with prolonged physical exertion, diabetes, or restricted food intake.

   a) If $[H^+]$ in the blood increases, in which direction will the equilibrium for the preceding reactions shift? Explain.

   b) How will this shift affect $CO_2$ levels? Explain.

   c) If breathing rate controls the amount of carbon dioxide exhaled as carbon dioxide gas, will breathing rate increase or decrease in metabolic acidosis in an attempt to return blood pH to appropriate levels? Explain.

20. Would hyperventilation (excessive breathing) result in respiratory acidosis or respiratory alkalosis? Justify your answer using your knowledge of the carbonic acid–bicarbonate buffer system and Le Châtelier's principle. Explain why fainting might be an appropriate body response to hyperventilation.

Amino acids and proteins can act as buffers in the body because they contain functional groups that act as weak acids or weak bases. For example, hemoglobin can help control blood pH during exercise by binding some of the excess protons generated in the muscles during anaerobic respiration.

21. Consider the side chain of the amino acid histidine shown here. The p$K_a$ of the protonated nitrogen in the side chain of free histidine is 6.0.

    a) Draw the structure of the conjugate base for the histidine side chain.

    b) Normal blood pH is tightly regulated between 7.35 and 7.45. Using pH 7.4 and the given p$K_a$ for histidine, determine the ratio of neutral histidine to protonated histidine for free histidine found in the blood.

    c) At pH 7.4, would histidine be more likely to act as an acid or a base? Explain.

22. Imagine you have two beakers filled with two different clear liquids. One is a buffer and the other is not. Design an experiment that would allow you to determine which beaker contains the buffer. Explain your experimental steps and the evidence you would use to draw your conclusion.

23. Explain what is wrong with the following statement: Buffers are neutral solutions.

Written by Heather Mernitz and Kim Schatz.

# Laboratory: How Does Cation Exchange Work?

- To understand and perform cation exchange in an ion exchange column

- To perform an acid–base titration

## INTRODUCTION

In this laboratory exercise, you will perform cation exchange, a process that happens constantly in soils. Soil particles are typically negatively charged, and positively charged cations in the soil are adsorbed onto them. These cations can exchange with other cations. This process is important for plants because they rely on soils to store cations via adsorption and release them through cation exchange. However, acid rain can change the environment of the soil in a manner that is detrimental to plants. For example, following an acid rain

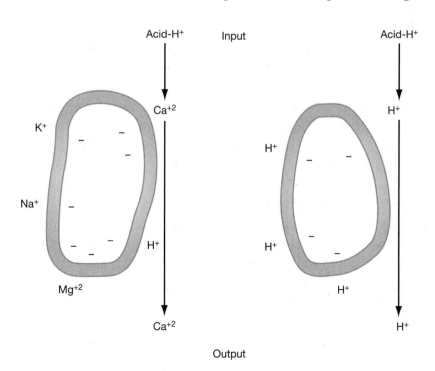

ChemConnections Activity Workbook

event, hydrogen ($H^+$) exchanges with nutrients such as calcium ($Ca^{2+}$) and potassium ($K^+$). These nutrients can then be leached away from the soil, leaving soils with predominantly $H^+$ cations.

In this laboratory exercise, you will determine the molecular weight of a sample using cation exchange followed by titration. You will be using a cation exchange resin in this experiment. This is a type of polymer (plastic) with attached sulfonic acid groups, $R-SO_3^-$, where R stands for part of the resin. Ions of opposite charge must also be present to ensure charge balance. These positive ions (cations) are not permanently bound to the resin and can be exchanged by appropriate adjustment of equilibrium conditions. You will be starting with the resin in the acid form, where the positive ion present is hydrogen. Addition of a copper solution then replaces the hydrogen ions with copper ions, freeing the hydrogen ions for analysis by titration with sodium hydroxide (NaOH).

$$2\ R-SO_3H(s) + Cu^{2+}(aq) \rightarrow (R-SO_3)_2Cu(s) + 2\ H^+(aq)$$

## PART I: CATION EXCHANGE PROCEDURE

1. Obtain approximately 10 mL of the ion exchange resin in a small beaker. If it is already in liquid, you will need to wait for it to settle to assure that you have approximately 10 mL. If your ion exchange resin is dry, add 20 mL of distilled water to the beaker. Place a tiny cotton wad at the base of a chromatography column and, using a funnel, pour the ion exchange resin into the column. Then drain the water into a waste beaker until the liquid is 0.5 cm above the resin. *Caution:* Never let the water level in the column drop below the resin level.

2. Obtain approximately 30 mL of 3.0 M HCl and add it to your chromatography column. This should be passed over the column at a rate no greater than 1 drop per second. 3.0 M HCl is a concentrated acid, so use all appropriate safety precautions, including safety goggles.

3. Once you have completed the acidification of the column, pass distilled water over the column until litmus paper indicates that the pH of the water going in matches that of the water coming out. When it does, your column is ready for ion exchange. (*Note:* Continue running water over the column until the pH matches).

4. Drain the water so that the level is 0.5 cm *above* the level of the resin bed. Wash your column by draining the rinse water as quickly as possible from the column.

5. During this process, have your partner obtain approximately 0.25 g of copper sulfate pentahydrate ($CuSO_4 \cdot 5H_2O$) and fully dissolve it in approximately 5 mL of distilled water. (*Note:* You do not need exactly 0.25 g of copper sulfate pentahydrate, you just need to know exactly how much you have.)

6. Add the copper solution to your prepared ion exchange column and let it flow through the column at a rate no greater than 1 drop per second into a clean, labeled Erlenmeyer flask until the liquid is again 0.5 cm above the level of the column bed. Record your observations.

7. At this point, the copper has stuck to the column and released protons ($H^+$). These protons can be rinsed at a much faster rate. Pass distilled water through the column until the outgoing water matches the pH of the incoming water. Collect all this

water in the same Erlenmeyer flask and set it to the side for titration during the next lab period.

8. Repeat the ion exchange procedure (starting at Step 5) two more times using your already prepared column on two separate copper sulfate samples. When you are finished, you should have three *labeled* Erlenmeyer flasks containing your samples. The column will need to be regenerated again (Steps 2 and 3) after three or four exchanges have been made on it, or if a color change caused by use reaches the bottom of the column.

9. Remove the copper from the column by adding 30 mL of 3 M HCl and running it through the column at 1 drop per second.

10. Stop here, and clean up. Take the ion exchange column apart. Pour the resin into the resin collection container and thoroughly rinse the ion exchange column.

## PART II: TITRATE THE COLLECTED HYDROGEN IONS WITH NaOH BY VOLUMETRIC TITRATION

11. Obtain approximately 50 mL of NaOH in a beaker and record its molarity.

12. Rinse the buret with approximately 10 mL of the NaOH solution and let the liquid drain through the buret tip into an empty waste beaker.

13. Position the buret in a double burette clamp on a ring stand.

14. Fill the buret with NaOH, being sure not to fill higher than the 0 mL line. Read the initial volume of the NaOH solution at the bottom of the meniscus. Your eye must be at the same level as the meniscus. *Write down the initial volume.*

15. Add two drops of phenolphthalein to one of your cation exchange samples.

16. Place the Erlenmeyer flask under the tip of the buret. A piece of white paper placed under the flask will make it easier to see the color changes.

17. While continuously swirling the flask to ensure thorough mixing, release the NaOH solution from the buret. Initially, a pink color will appear at the point where the NaOH comes in contact with the solution in the flask; however, this color disappears quickly. As the endpoint nears, the color will disappear more slowly.

18. NaOH should be added drop by drop until one drop turns the entire solution in the flask pink. This pink color should remain at least 15 seconds while the solution is being swirled. This is the endpoint. If you overshoot the endpoint, you will not be able to count that sample. When you have reached a satisfactory endpoint, *read the final volume of the buret.*

19. Repeat steps 14–18 for your other two samples.

## PART III: CALCULATIONS

20. For each of the three samples, use titration calculations to calculate the number of moles of hydrogen ion in the sample.

21. For each of the three samples, use the balanced equation to convert the number of moles of hydrogen ion to the number of moles of copper ion. How many moles of $CuSO_4 \cdot 5H_2O$ does this correspond to?

22. For each of the three samples, use the moles of $CuSO_4 \cdot 5H_2O$ from Question 2 and the mass of sample to calculate the molecular weight of your sample.

23. For each of the three samples, calculate the percent error for the molecular weight of $CuSO_4 \cdot 5H_2O$. You will do this as follows:

a) Use the atomic masses from the periodic table to calculate the actual molecular weight of $CuSO_4 \cdot 5H_2O$.

b) Calculate percent error using the following formula:

$$\text{percent error} = \left( \frac{\text{experimental MW} - \text{theoretical MW}}{\text{theoretical MW}} \right) \times 100\%$$

Adapted by Sharon Anthony and George Lisensky from the ChemConnections module "Soil Equilibria: What Happens to Acid Rain?" by George Lisensky, Roxanne Hulet, Michael Beug, and Sharon Anthony.

# What Controls the Properties of Elements?

■ **To develop your analytical skills in reading and summarizing graphical data**

■ **To explore periodic properties of the elements**

■ **To apply your knowledge of periodic trends to predict relative properties of the elements**

## INTRODUCTION

The periodic properties of the elements depend on the fact that the elements in the last group (Group 18, the noble gases) have extraordinary chemical stability. These elements have very stable electron configurations in filled electron shells. The electrons in a noble gas have little tendency to be lost, gained, or shared. Atoms of other elements can achieve this stability, but they must either gain or lose electrons, or share electrons in covalent bonds, to have filled electron shells.

Ionization energy is the energy required to remove one electron from an atom or an ion. $M$ in the following equation can be any element.

$$M \rightarrow M^+ + e^-$$

The energy required to remove the first electron is called the first ionization energy. The energy required to remove a second electron is called the second ionization energy.

$$M^+ \rightarrow M^{2+} + e^-$$

Likewise, the energy required to remove a third electron is called the third ionization energy. The first electron affinity is the energy associated with removing one electron from the singly charged negative ion.

$$M^- \rightarrow M + e^-$$

The larger the electron affinity, the greater the attraction the neutral atom has for an additional electron. The noble gases will have very high first ionization energy and very low

**265**

electron affinity because neutral atoms of these elements are very stable with their filled electron shells and don't have a tendency to gain or lose extra electrons.

Many of the periodic properties of elements can be correlated with the sizes of the atoms. Electrons fill in the shell across a row of the periodic table and, in general, size decreases from left to right as the negative electrons are more strongly attracted to the increasingly positive nucleus. Going down a column in the periodic table, electrons fill shells that are farther away from the nucleus and, in general, size increases down a group of the periodic table. When atoms form ions, the gain of electrons leads to an increase in radius and the loss of electrons leads to a reduction in radius, both due to the perturbation of the proton-electron balance.

Size and charge determine how well an atom attracts its own or another atom's electrons. This property is called electronegativity. Higher electronegativities arise from higher nuclear charges (atomic numbers) and smaller atomic sizes. This results in generally increasing electronegativities going from left to right across a row in the periodic table. Electronegativity generally increases from bottom to top within a group.

## PART I: EXAMINING DATA

This activity contains figures of a variety of elemental properties. In each case, a stick plot compares changes as a function of atomic number, and a three-dimensional periodic table (where the vertical height corresponds to the magnitude of the property) facilitates comparisons across rows and down columns. Refer to these figures to answer the questions that follow about the periodic trends you see. For example, for which elements does the property trend reach local maxima?

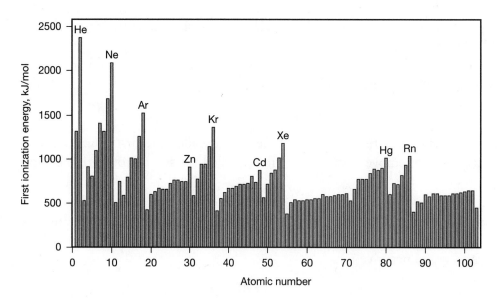

First ionization energy as a function of atomic number.

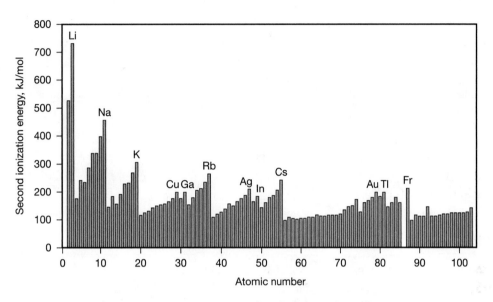

Second ionization energy as a function of atomic number.

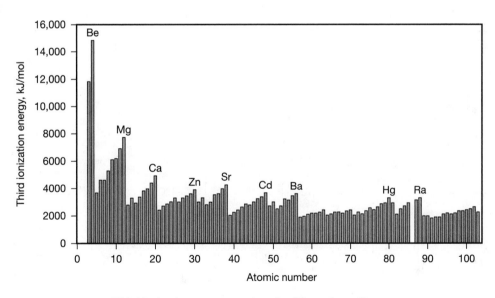

Third ionization energy as a function of atomic number.

## FIRST IONIZATION ENERGY

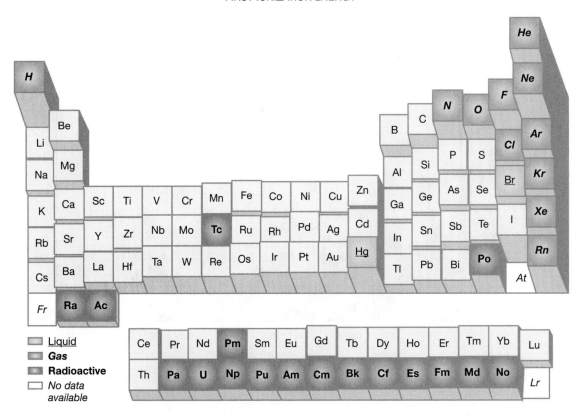

First ionization energy superimposed on the periodic table.

## SECOND IONIZATION ENERGY

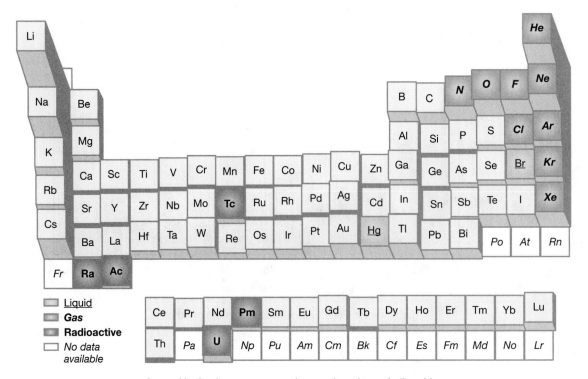

Second ionization energy superimposed on the periodic table.

THIRD IONIZATION ENERGY

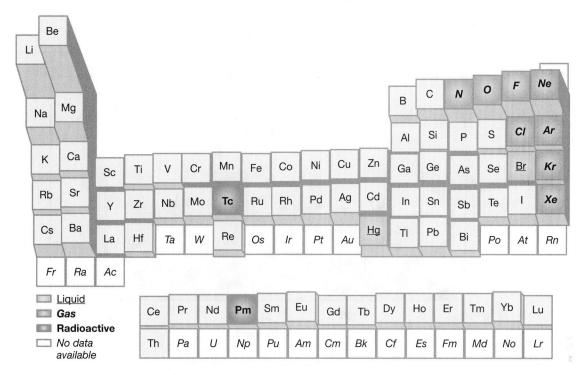

Third ionization energy superimposed on the periodic table.

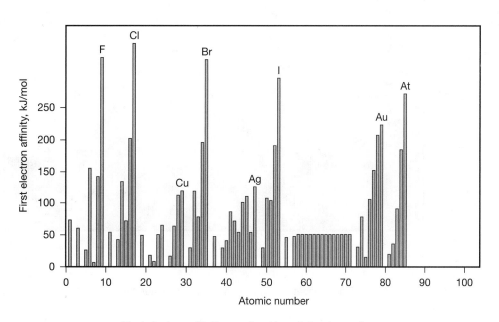

First electron affinity as a function of atomic number.

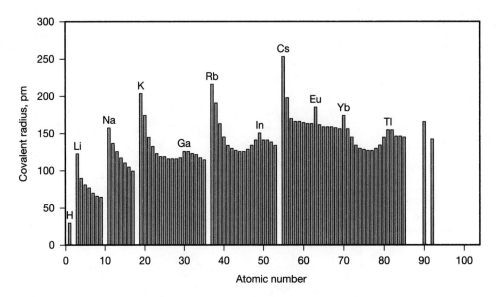

Covalent radius as a function of atomic number.

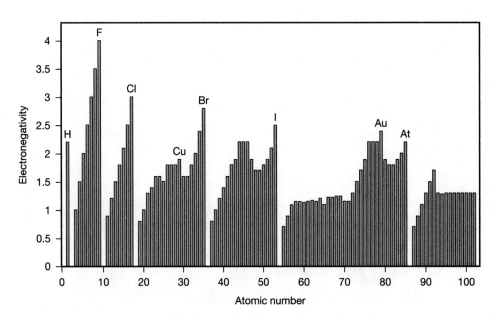

Electronegativity as a function of atomic number.

## ELECTRON AFFINITY

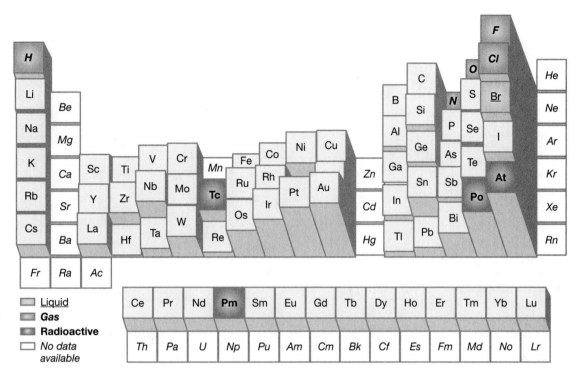

Electron affinity superimposed on atomic number.

## COVALENT RADIUS

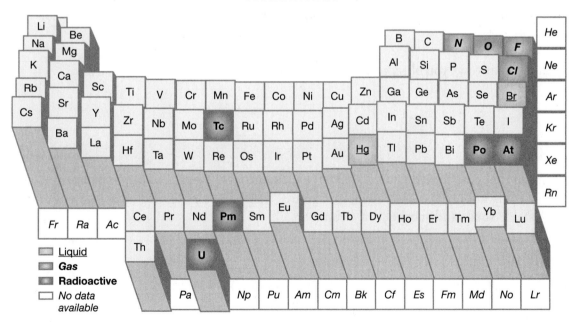

Covalent radius superimposed on the periodic table.

ELECTRONEGATIVITY

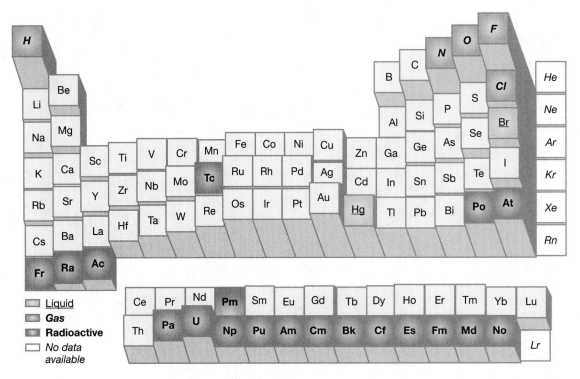

Electronegativity superimposed on the periodic table.

1. On the periodic table on the next page, shade the elements at the peaks of the first ionization energies one color, those at the peaks of the second ionization energies a second color, and those at the peaks of the third ionization energies a third color.

2. Explain what you observe. Are you observing periodic behavior?

3. On the same periodic table, shade the elements at the peaks of the first electron affinity a fourth color. How do the groups shaded so far correlate with the notion of stable shells and subshells of electrons?

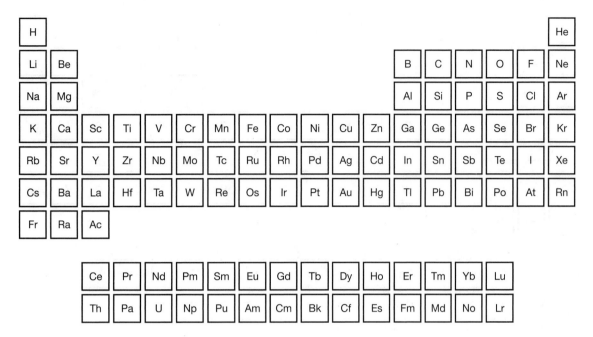

Periodic table for Questions 1 through 3.

4. On the periodic table on the next page, shade the elements at the peaks of covalent radius. How do the elements you just shaded compare with those shaded for the ionization energies or electron affinity? Why is this?

5. On the same periodic table, use a different color to shade the elements at the peaks of electronegativity for each period. Compare the elements you just shaded with those shaded for ionization energy, electron affinity, and the covalent radius. What conclusions can you draw about the relationships among these three periodic properties? Give reasons for your conclusions.

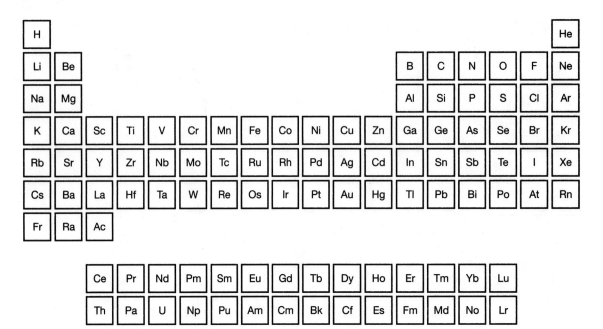

Periodic table for Questions 4 through 5.

6. The periodic properties of the elements may differ in some respects from what you would expect. Did you note any such differences or discrepancies in Part I of the worksheet?

## PART II: PRACTICE QUESTIONS

7. Arrange the elements in order from smallest to largest radius.

   a) Al, B, C, K, Na

   b) Ca, Ge, P, Rb, Sr

8. Circle the element with largest radius in each pair.

   a) Al, N

   b) In, Sn

   c) Cs, Rb

   d) As, Br

   e) Cl, Cl$^-$

   f) O, O$^{2-}$

   g) Na, Na$^+$

9. Arrange the elements in order from smallest to largest first ionization energy.

   a) Al, F, Mg, P

   b) C, K, Li, N

## References

Atomic mass values for naturally occurring isotopic mixtures from *Pure and Applied Chemistry* **1983,** *55,* p 1101.

Ionization energies from Moore, C. E. *NBS Circular* **1970,** *34 and 467*; Martin, W. C., et al. *Journal of Physical Chemistry Reference Data* **1974,** *3,* p 771; Katz, J. J.; Seaborg, G. T.; Morss, L. R. *The Chemistry of the Actinide Elements,* 2nd ed.; Chapman and Hall: London, 1986.

Electron affinity values from Hotop, H.; Lineberger, W. C. *Journal of Physical Chemistry Reference Data* **1985,** *14,* p 731; Pearson, R. G. *Inorganic Chemistry* **1991,** *30,* pp 2856–2858; Miller, T. M.; Miller, A. E. S.; Lineberger, W. C. *Physics Review A* **1986,** *33,* pp 3358–3359.

Covalent radii from Dean, J. A. *Lange's Handbook of Chemistry*; McGraw–Hill: New York, 1985.

Electronegativity values from Pauling, L. *Chemical Bond*; Cornell University Press: Ithaca, NY, 1967.

Adapted by George Lisensky from the ChemConnections module "Build a Better CD Player: How Can You Get Blue Light from a Solid?" by George C. Lisensky, Herbert Beall, Arthur B. Ellis, Dean J. Campbell, and Joanne Stewart.

# What Are the Building Blocks of Life?

- To review the key functional groups and common structural features present in carbohydrates, lipids, proteins, and nucleic acids

- To identify classes of biomolecules in monomeric and polymeric form

## INTRODUCTION

Biomolecules are those mainly organic molecules that are involved in the everyday function of living systems. These molecules are often very large and complex. To understand these molecules and the role they play in living systems, you can "break" them down into smaller pieces, units, or individual functional groups so that their physical properties and biological activity are easier to see and understand.

## PART I: THE BUILDING BLOCKS OF BIOMOLECULES

Carbohydrates, lipids, proteins, and nucleic acids are four important classes of biomolecules. The structural features (i.e., functional groups, shape, and size of nonpolar regions) lead to the observed reactions and functions of these molecules. Here are some structures that represent the monomers (the smallest repeating unit) of carbohydrates, lipids, proteins, and nucleic acids.

## Carbohydrates: Monosaccharides (1 Sugar Unit)

Selected monosaccharides

1. What characteristic functional groups do you see in the monosaccharides in the preceding figure?

2. Do monosaccharides have any similarities in structure other than their functional groups? (If you were given a picture of a biomolecule, how would you identify it as a carbohydrate?)

3. Are monosaccharides likely to be soluble in water? Why or why not?

4. Identify the types of intermolecular forces you would expect to see between adjacent monosaccharides.

**Lipids: Fatty Acids**

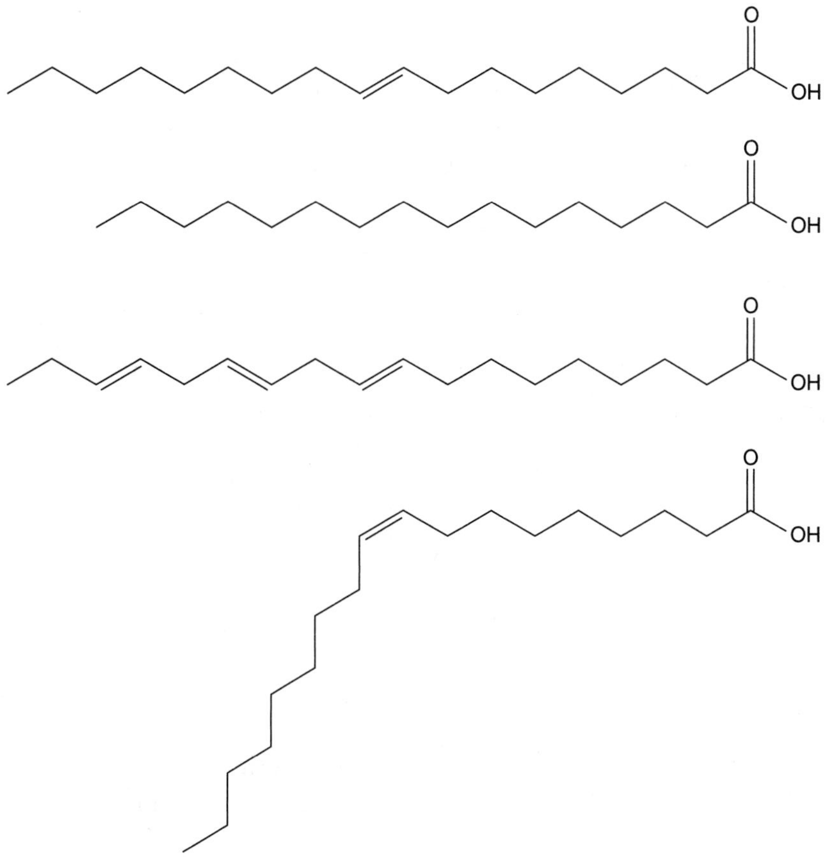

Selected fatty acids

5. What characteristic functional groups do you see in the fatty acids in the preceding figure?

6. Do fatty acids have any similarities in structure other than their functional groups? (If you were given a picture of biomolecule, how would you identify it as a lipid?)

7. Are fatty acids likely to be soluble in water? Why or why not?

8. Identify the types of intermolecular forces you would expect to see between adjacent fatty acids.

## Proteins: Amino Acids

Selected amino acids

9. What characteristic functional groups do you see in the amino acids in the preceding figure? What functional groups are unique to individual amino acids?

10. Do amino acids have any similarities in structure other than their functional groups? (If you were given a picture of biomolecule, how would you identify it as a protein?)

11. Are amino acids likely to be soluble in water? Why or why not?

12. Identify the types of intermolecular forces you would expect to see between adjacent amino acids.

## Nucleic Acids (DNA, RNA): Nucleotides

Selected nucleotides

13. Which characteristic functional groups do you see in the nucleotides in the preceding figure?

14. Do nucleotides have any similarities in structure? (If you were given a picture of biomolecule, how would you identify it as a nucleic acid?)

15. How would you distinguish between a nucleic acid and a carbohydrate?

16. Are nucleotides likely to be soluble in water? Why or why not?

17. Identify the types of intermolecular forces you would expect to see between adjacent nucleotides.

## PART II: BIOPOLYMERS: CONNECT THE BLOCKS

The monomers you just analyzed can be covalently linked to form larger biomolecules. Two to three fatty acids come together to form triglycerides, a class of lipids. Hundreds to thousands of amino acids and nucleotides come together to form proteins and nucleic acids, respectively. Thousands to tens of thousands of monosaccharides can come together to form complex carbohydrates.

18. Using the patterns and properties you inferred in Part I, identify the polymers here by biomolecule class: carbohydrate, lipid, protein, or nucleic acid.

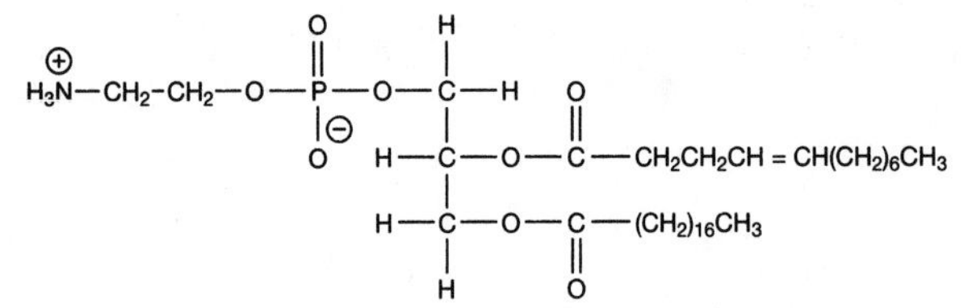

a) _____

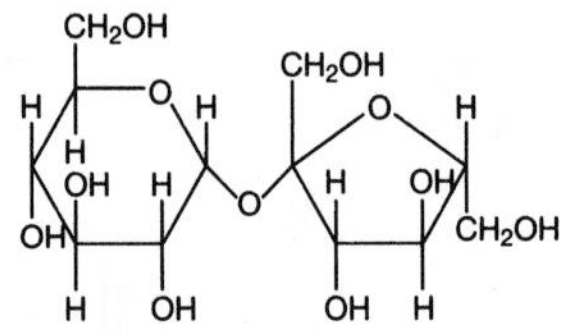

b) _____

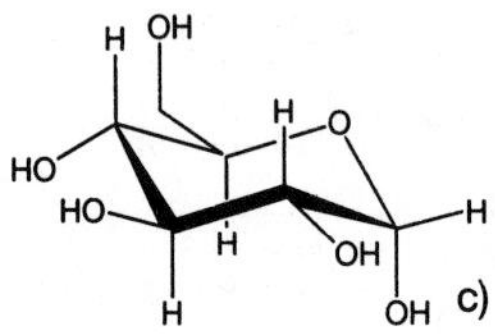

c) _____

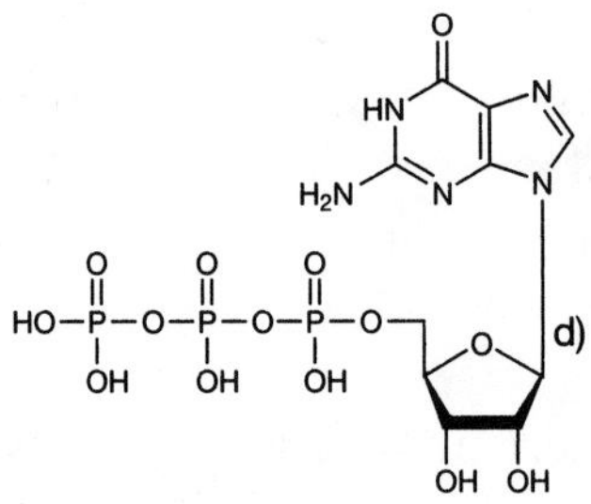

d) _____

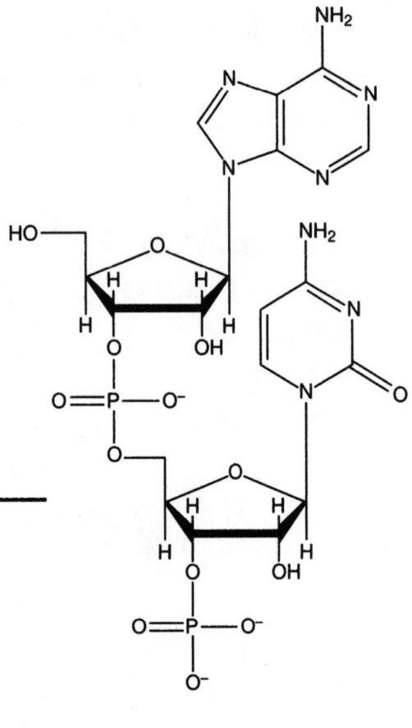

e) _____

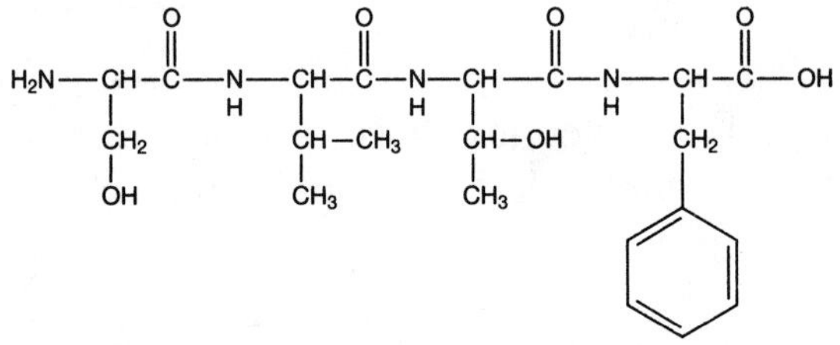

f) _____

## PART III: REACTIONS OF BIOMOLECULES

Now we are going to take a closer look at the reactions involved in building biomolecule polymers from their monomers. These reactions require both monomers as building blocks and chemical energy in the form of adenosine triphosphate (ATP).

19. Table sugar, sucrose, is a disaccharide made from a reaction that combines glucose and fructose. Adding further monosaccharides to this chain will result in polysaccharides, which can grow very large. Glycogen, the storage form for glucose in the muscle of animals, can be composed of tens of thousands of glucose monomers. Plants store glucose in long polymers called starches.

Formation of maltose from two glucose molecules

a) Identify the functional groups in the reactants and the functional groups in the products. What functional groups are lost in the reaction and what new functional groups are formed?

b) The bond that links the two glucose molecules together to form maltose is called a glycosidic linkage. What specific functional group is involved in this bond?

c) How will the new chemical functional group affect the way water interacts with the disaccharide compared to the individual monosaccharides?

d) How do you think the physical properties (solubility, melting point) of disaccharides will be different from monosaccharides?

20. A triglyceride is a lipid comprised of three fatty acids attached to a glycerol backbone. Triglycerides are found in our diets in the vegetable oils and animal fats we consume. Because triglycerides contain more than twice the energy per gram as carbohydrates and proteins, they are a concentrated source of energy for our bodies and an efficient storage form of energy.

   High levels of triglycerides in the bloodstream have been associated with heart disease and stroke, especially when coupled with an unfavorable ratio of low-density lipoprotein to high-density lipoprotein (LDL/HDL) cholesterol.

Formation of a triglyceride from glycerol and fatty acids

   a) Identify the functional groups in the reactants and the functional groups in the products. What functional groups are lost in the reaction and what new functional groups are formed?

   b) How will the new chemical functional groups affect the way water interacts with the triglyceride as opposed to how it interacts with the individual fatty acids?

21. Proteins are long strands of amino acids. They serve many roles in the body, including structure (collagen), protection (immunoglobulin), hormones (insulin), transport (hemoglobin), storage (myoglobin), motility (myosin), and catalysis (enzymes). Regardless of their purpose, a common feature of proteins is that their three-dimensional shape is intimately connected to their function. When a protein loses its shape (called denaturation), its physical and chemical properties are profoundly changed and it is no longer able to function appropriately. You have probably observed protein denaturation when frying an egg. The heat causes the proteins in the egg white to unfold, which results in observable changes in the egg's appearance and taste.

Formation of a dipeptide from two amino acids

a) Identify the functional groups in the reactants and the functional groups in the products. What functional groups are lost in the reaction and what new functional group is formed?

b) The bond that links the two amino acids together to form a dipeptide is called a *peptide bond.* What specific functional group is involved in this bond?

c) How will the new chemical functional group affect the way water interacts with the dipeptide as opposed to how it interacts with the individual amino acids?

22. Nucleotides consist of a sugar, a nitrogenous base, and one or more phosphate groups. These nucleotides can be linked together to form nucleic acids. If the sugar is ribose and the bases are adenine, guanine, cytosine, and uracil, the nucleic acid is called ribonucleic acid (RNA). If the sugar is deoxyribose (which lacks the hydroxyl group on carbon 2) and the bases are adenine, guanine, cytosine, and thymine, the nucleic acid is called deoxyribonucleic acid (DNA). All the information required to assemble a living organism is encoded in that organism's DNA. Several types of RNA exist, all of which play roles in the use of the information stored in the DNA, including the synthesis of proteins.

a) Identify the functional groups in the reactants and the functional groups in the products. What functional groups are lost in the reaction and what new functional groups are formed?

Formation of a dinucleotide from two mononucleotides

b) How will the new chemical functional group affect the way water interacts with the product as opposed to how it interacts with the individual nucleotides?

23. What similarities or patterns do you see in reactions in problems 19–22? What differences do you see for the individual classes of biomolecules?

24. There is a second product that is not shown in any of the preceding reactions. What is this second product? To what class of organic reactions do each of these anabolic reactions belong?

25. Metabolism is the balance between two opposing types of processes. You have already worked with anabolic processes, in which biomolecules are created from simple monomers. Catabolic processes break down larger biomolecules into their simpler components. Answer the following questions for the biomolecule below.

a) What class of biomolecules is represented in the preceding figure? What are the monomers for this class of biomolecules called? How many of these monomers are present in the molecule?

b) You have already built an example of this class of biomolecules from two monomers. Reverse this process by breaking the linkage between monomers. This catabolic reaction is an example of hydrolysis; you will need to add one water molecule for each linkage broken.

26. A cell membrane forms a barrier between a cell's contents and the external environment. The membrane itself consists of a phospholipid bilayer. The assembly of a bilayer does not require a chemical reaction. To get an idea of how a bilayer forms, you can add oil droplets to a small dish of water. You should observe that the oil droplets aggregate spontaneously.

a) Below is the structure of a phospholipid. Circle the part of the molecule that is most polar (or hydrophilic). Box the part of the molecule that is most nonpolar (or hydrophobic).

Phosphatidylethanolamine

b) You may be familiar with the cartoon sketch of a phospholipid, shown as a head with two tails:

What exactly is the head? Is the head polar or nonpolar? What exactly are the tails? Are the tails polar or nonpolar?

c) There are three different ways by which the nonpolar parts of a phospholipid may hide from water.

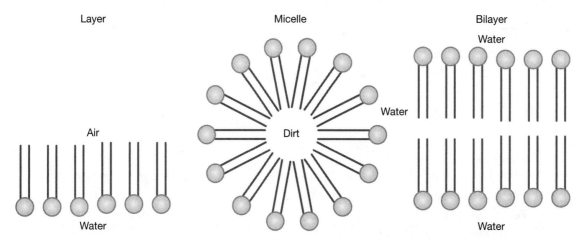

What regions are hydrophobic and what regions are hydrophilic in each sketch in the preceding figure? Describe the forces that account for the three arrangements.

27. Integral membrane proteins, such as the ion channel shown below, sit within the cell membrane such that a portion of the protein associates with the interior of the bilayer (tails) and another portion of the protein associates with the exterior of the bilayer (heads). Describe the chemical character of the amino acid side chains you might find on the exterior of this membrane protein at each of these two locations.

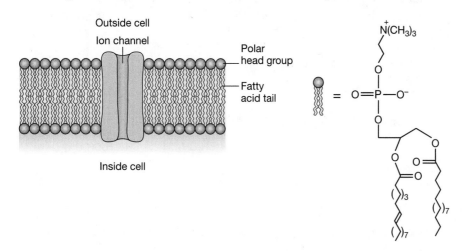

Cartoon depiction of a cellular membrane with an embedded integral membrane protein.

Written by Theodore Gries and Heather Mernitz.

# How Does a Nuclear Reactor Work?

■ To apply basic knowledge of atomic structure and unit conversions to nuclear reactions

■ To learn how a nuclear reactor functions and compares to fossil-fuel sources of electrical energy

## INTRODUCTION

When it comes to climate change, how to meet the world's growing electrical energy needs in a carbon neutral manner has taken center stage. Nuclear energy has been grouped with solar and wind energies as possible zero-carbon-emission sources. To better understand the inclusion of nuclear energy in this classification, you will investigate how a nuclear reactor functions and make comparisons to coal electrical generation to evaluate environmental impact.

1. The Kewaunee Nuclear Power Station in Carlton, Wisconsin, contains one nuclear reactor unit. While operating normally, this unit has a gross electrical generating output of approximately 556 MW. Determine the joules per second of electrical energy generated by this unit. (*Hint:* M = mega ($10^6$); W = watt = joule per second.)

2. The movement of a wire through a magnetic field generates electrical current (net unidirectional movement of electrons) within the wire. In a large-scale nuclear generator, the kinetic energy of steam drives the rotation of a turbine, which rotates a magnet (and associated magnetic field) inside a coil of wire. Assuming 100% efficiency in the conversion of kinetic energy from steam into electrical energy, how

much kinetic energy of steam is required each second to turn the turbine at the Kewaunee Nuclear Station (in J/s)?

3. In current nuclear reactors, the energy needed to generate steam comes from the fission (breaking apart) of large, unstable nuclei. Fill in the blanks for the following example reaction that might occur in a nuclear reactor:

$$n + {}^{235}_{92}U \rightarrow {}^{236}_{92}U \rightarrow {}^{89}_{36}\underline{\quad} + {}^{144}_{56}\underline{\quad} + \underline{\quad}n + \underline{\quad}p^+$$

4. To be an effective energy source, a nuclear reaction must sustain itself. What characteristic of the preceding reaction allows it to continue without input from an external source?

5. When one ${}^{235}_{92}U$ atom undergoes the nuclear fission described in Question 3, approximately $3.244 \times 10^{-11}$ J of energy is released. Assuming that 100% of the energy released by each atomic fission event is converted into heat to generate steam in the reactor's boiler, how many ${}^{235}_{92}U$ atoms must split each second to power the Kewaunee unit?

6. How many grams per year of ${}^{235}_{92}U$ are required to power the Kewaunee unit?

7. The total mass of the products of the fission reaction is actually slightly less than the total mass of the reactants. The conversion of this mass into energy is the source of the heat energy. Using Einstein's famous equation, $E = mc^2$, calculate how much mass in grams are converted into energy each year to power the Kewaunee unit. (*Hint:* Joule $= \dfrac{m^2\,kg}{s^2}$.)

8. What percentage of the total mass required to power the Kewaunee unit is converted into energy?

9. In reality, the conversion of energy from one form to another is never 100% efficient. Nuclear reactors average 35% efficiency in conversion of fission energy to electricity. Realistically, how many grams per year of $^{235}_{92}U$ are required to power the Kewaunee unit?

10. In the United States in 2009, approximately 7460 J of electrical energy were produced for every gram of coal burned. How many grams per year of coal would be required to create the same amount of power as the Kewaunee unit?

11. How do the masses of uranium and coal in Questions 9 and 10 compare?

12. What are other sources of carbon emissions associated with coal-fired electrical generation besides the actual combustion reaction?

13. Are there any carbon-emitting processes associated with using nuclear reactions to generate electricity?

14. Is nuclear power truly carbon-emission neutral?

Written by Theodore Gries.

# When Did You Live?

## INTRODUCTION

Accurately determining the age of an artifact is an important aspect in modern archeology. Our understanding of radioactive decay and the interpretation of concentrations of various isotopes in our environment, as compared with aged objects, have greatly advanced our ability to determine an object's age independent of historical (written and oral) records. In this activity, you will examine how to determine the age of an object from the perspective of a future researcher looking at artifacts that you leave behind today.

1. Cotton is 44.4% carbon by mass, and the average cotton t-shirt is 300 grams. Under current environmental conditions, $6.00 \times 10^{-12}$ Ci (curies) of $^{14}_{6}C$ are deposited in an organic object per gram of total carbon. How many curies of $^{14}_{6}C$ are in a t-shirt that you have recently worn?

2. Of the total carbon present in your t-shirt, 99.9% by mass is $^{12}_{6}C$. What is the mass of $^{12}_{6}C$ in your t-shirt?

ChemConnections Activity Workbook

3. The half-life of $^{14}_{6}C$ is 5700 years. If a researcher in the year 11,510 were to find your preserved t-shirt, how many curies of $^{14}_{6}C$ would be present in it?

4. If a researcher were to find your preserved t-shirt to have $5.00 \times 10^{-10}$ Ci of $^{14}_{6}C$, approximately what year is the researcher performing the analysis?

5. How much $^{12}_{6}C$ of your original t-shirt would remain in each of the future years from Questions 3 and 4 if it is very well preserved?

6. Imagine that 2 million years were to pass and *Homo sapiens* were to evolve into a new species. This new species finds the remains of your t-shirt encased in a cavity within a calcite rock. How many curies of $^{14}_{6}C$ would be present in the residue of the cavity?

7. Is $^{14}_{6}C$ dating a viable technique for this new species to study our time period? Why or why not?

8. To study longer time intervals, other isotopes with considerably longer half-lives may be used. Fill in the blanks in the following radioactive decay reactions.

$$^{238}_{92}U \xrightarrow{4.5\times10^9 \text{ years}} \underline{\quad} + {}^4_2He \xrightarrow{24.1 \text{ days}} {}^{234}_{91}Pa + \underline{\quad} \xrightarrow{1.18 \text{ month}} \underline{\quad} + {}^0_{-1}e \xrightarrow{2.49\times10^5 \text{ years}} {}^{230}_{90}Th + \underline{\quad}$$

9. $^{238}_{92}U$ is incorporated into calcite rock whenever it is deposited. Which isotopes from Question 8 would you quantify if you were designing an experiment 2 million years in the future to determine approximately when your t-shirt was encapsulated in calcite?

## References

Question 1, Deposition of $^{14}_{6}C$ in an organic object per gram of total carbon: Argonne National Laboratory.

Questions 8 and 9, radioactive decay reactions: Schwarcz, H. P. *Philosophical Transactions of the Royal Society of London* **1992**, *337*, pp 131–137.

Written by Theodore Gries.

# What Do the Data Tell Us about Climate Change?

- **To develop your analytical skills in reading and summarizing graphical data, trends, and scientific confidence**

## INTRODUCTION

The United Nations brought together the Intergovernmental Panel on Climate Change (IPCC) to study climate change and to inform the global community about the potential consequences of climate change. The IPCC comprises hundreds of the most respected climate experts in the world. They study historical and current climate data and create models to predict future consequences. The following worksheet refers to data and figures taken from the IPCC report *Climate Change 2007: The Scientific Basis*. Completing this worksheet will help you to better understand how scientific data and uncertainty are represented in graphical forms.

## PROBLEMS

1. Examine the graph given here and read the associated caption.

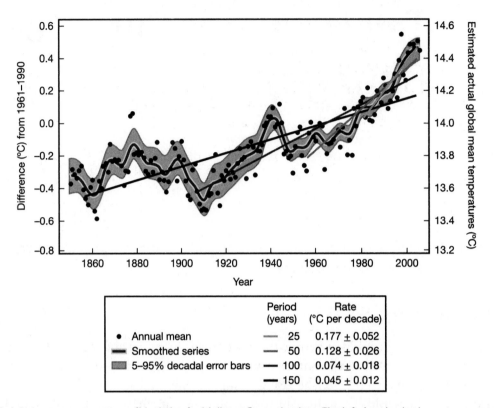

| | | Period (years) | Rate (°C per decade) |
|---|---|---|---|
| ● | Annual mean | — 25 | 0.177 ± 0.052 |
| ▬ | Smoothed series | — 50 | 0.128 ± 0.026 |
| ▓ | 5–95% decadal error bars | — 100 | 0.074 ± 0.018 |
| | | — 150 | 0.045 ± 0.012 |

Annual global mean temperatures (black dots) with linear fits to the data. The left-hand axis shows temperature anomalies relative to the 1961–1990 average, and the right-hand axis shows estimated actual temperatures, both in degrees Celsius. Linear trends are shown for the last 25 (yellow), 50 (orange), 100 (purple), and 150 years (red). The smooth blue curve shows decadal variations with the decadal 90% error range shown as a pale blue band about that line. The total temperature increase from the period 1850–1899 to the period 2001–2005 is 0.76°C ± 0.19°C.
A color version of this figure can be found at www.ipcc.ch/publications_and_data/ar4/wg1/en/tssts-3-1-1.html

a) Is the IPCC reporting that the temperature has increased over the last 150 years? If so, by how much?

b) Describe how the rate of temperature increase has changed over this time period?

2. Examine the figure here and carefully read its caption.

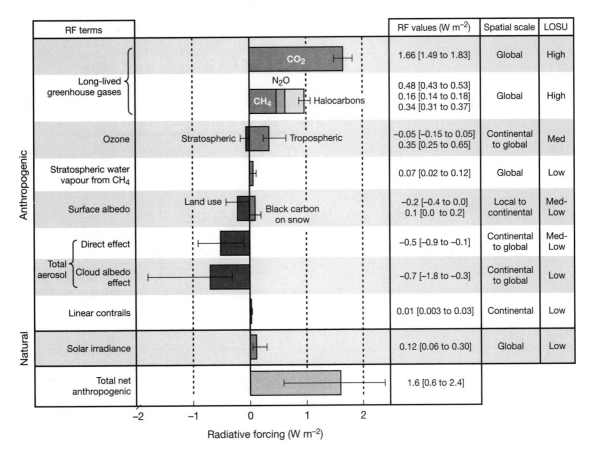

Global mean radiative forcings (RF) and their 90% confidence intervals in 2005 for various agents and mechanisms. Columns on the right-hand side specify best estimates and confidence intervals (RF values); typical geographical extent of the forcing (Spatial scale); and level of scientific understanding (LOSU) indicating the scientific confidence level. Errors for $CH_4$, $N_2O$ and halocarbons have been combined. The net anthropogenic radiative forcing and its range are also shown. Best estimates and uncertainty ranges can not be obtained by direct addition of individual terms because of the asymmetric uncertainty ranges for some factors; the values given here were obtained from a Monte Carlo technique. Additional forcing factors not included here are considered to have a very low LOSU. Volcanic aerosols contribute an additional form of natural forcing but are not included because of their episodic nature. The range for linear contrails does not include other possible effects of aviation on cloudiness. A color version of this figure can be found at www.ipcc.ch/publications_and_data/ar4/wg1/en/spmsspm-human-and.html

a) Radiative forcing has units of watts per square meter. What does it mean for something to have a positive radiative forcing? A negative radiative forcing?

b) Which of the factors shown in the preceding graph have a warming influence on climate? Which of these factors are anthropogenic (human caused) in origin?

    c) Which single factor has the most influence on global warming?

    d) Which factors have a cooling influence? Which of these factors are anthropogenic in origin?

    e) Summarize how certain the IPCC scientists are about the different radiative forcings. Which radiative forcings are well understood? Which have considerable uncertainty?

3. Summarize each graph shown on the following page. Consider concentration units, time scale, past trends, and recent trends (shown as an inset in the larger graph).

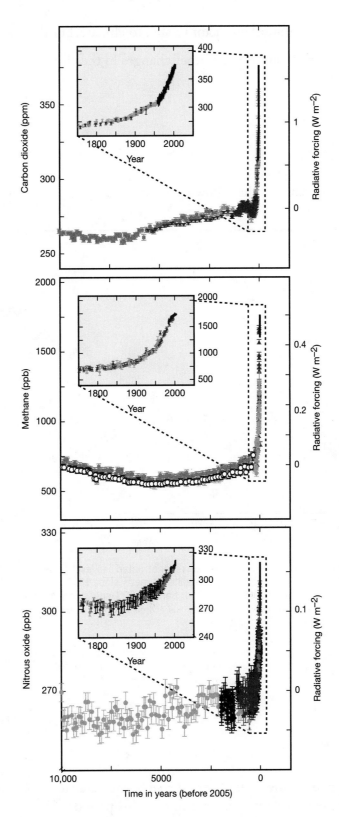

Atmospheric concentrations of carbon dioxide, methane, and nitrous oxide over the preceding 10,000 years (large panels) and since 1750 (inset panels). Measurements are shown from ice cores (symbols with different colors for different studies) and atmospheric samples (solid red lines). The corresponding radiative forcings are shown on the right-hand axes of the large panels.

A color version of this figure can be found at www.ipcc.ch/publications_and_data/ar4/wg1/en/spmsspm-human-and.html

4. Temperature is one important factor related to climate, but there are many others.

a) Aside from temperature, what other changes in the climate system do you anticipate?

b) Describe the trends in the following figure.

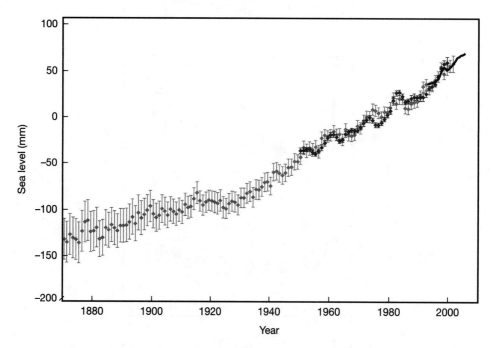

Annual averages of the global mean sea level based on reconstructed sea level fields since 1870 (light gray), tide gauge measurements since 1950 (dark gray), and satellite altimetry since 1992 (solid black line). Units are in millimeters relative to the average for 1961–1990. Error bars are 90% confidence intervals.

A color version of this figure can be found at www.ipcc.ch/publications_and_data/ar4/wg1/en/tssts-3-3-3.html

c) What happens to the level of confidence in the data from 1870 to 2000? Why do you think the uncertainty changes during this period?

5. Examine the graph here and read its caption carefully. According to the climate models used by the IPCC, can natural influences alone explain the temperature changes we have seen in the past 100 years?

GLOBAL

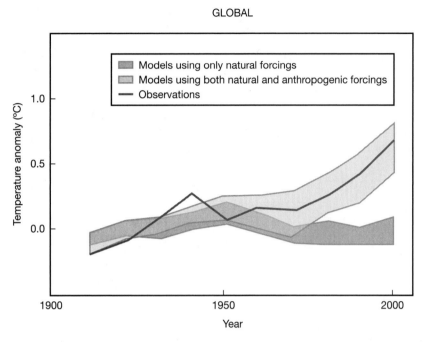

Comparison of observed global-scale changes in surface temperature with results simulated by climate models using natural and anthropogenic forcings. Decadal averages of observations are shown for the period 1906–2005 (black line) plotted against the center of the decade and relative to the corresponding average for 1901–1950. Darker blue-shaded bands show the 5%–95% range for 19 simulations from five climate models using only the natural forcings due to solar activity and volcanoes. Lighter red-shaded bands show the 5%–95% range for 58 simulations from 14 climate models using both natural and anthropogenic forcings.

A color version of this figure can be found at www.ipcc.ch/publications_and_data/ar4/wg1/en/spmsspm-understanding-and.html

---

Adapted from the ChemConnections module "What Should We Do about Global Warming?" by Sharon Anthony, Tricia A. Ferrett, and Jade Bender.

---

# Laboratory: How Is Copper Extracted from Ore?

- To learn how surfactants are used to separate copper-rich particles from other particles

- To explore the chemical equations involved in the pyrometallurgical and hydrometallurgical purification methods

- To use thermodynamic data ($\Delta G$ and $K_{sp}$) to predict the feasibility of reactions in copper processing

- Explain the role of bacteria in the processing of copper ores

## INTRODUCTION

Copper-containing minerals appear in nature in many different forms. The table here shows several of the more common minerals. The extraction of copper from ore is performed in one of two ways, pyrometallurgy or hydrometallurgy, depending on the type of mineral(s) present in the ore.

| COPPER MINERALS | | | |
|---|---|---|---|
| Mineral | Formula | Mineral | Formula |
| Antlerite | $CuSO_4 \cdot 2\,Cu(OH)_2$ | Chrysocolla | $CuSiO_3 \cdot 2\,H_2O$ |
| Azurite | $2\,CuCO_3 \cdot Cu(OH)_2$ | Covellite | $CuS$ |
| Bornite | $Cu_5FeS_4$ | Cuprite | $Cu_2O$ |
| Brochantite | $CuSO_4 \cdot 3\,Cu(OH)_2$ | Malachite | $CuCO_3 \cdot Cu(OH)_2$ |
| Chalcopyrite | $CuFeS_2$ | Native Copper | $Cu$ |
| Chalococite | $Cu_2S$ | Tenorite | $CuO$ |

About 90% of the world's copper ore deposits contain primarily sulfidic minerals. The percentage of copper by mass in these deposits is fairly low, between 0.5% and 2%. The methods by which this copper ore is converted to copper metal are generally referred to as *smelting* and *pyrometallurgy*. Smelting takes the impure ore and reduces the copper ion into an impure metal that is then purified electrochemically. During the smelting process, sulfur and iron ions are oxidized.

## PYROMETALLURGY

To effectively smelt an ore containing 0.5% to 2% copper by mass, it is necessary to first concentrate the ore. This copper enrichment is done by first crushing the ore to small particles and then separating the copper-containing pieces through a process called *froth flotation*.

The crushing is performed in three steps. The primary crusher takes rocks from the mine and breaks them into smaller rocks, about 200 cm in diameter. This initial crushing is followed by two successive grinding steps, which reduce the particles to ~3 cm and ~0.3 cm, respectively. The final size required depends on the size of the copper ore grains in the rock—it is most efficient to crush the rock small enough so that each particle contains a single mineral grain.

After the minerals are broken into small particles, the particles containing copper sulfides are separated from the other particles. This is accomplished by treating an aqueous slurry with a surfactant that preferentially adheres to the surface of the copper sulfides and renders these particles hydrophobic. The most commonly used surfactants are the xanthates, examples of which are shown here.

Potassium amyl xanthate          Sodium isopropyl xanthate

The polar head group, drawn toward the left, adheres to the copper sulfide mineral and the hydrophobic chain extends away from the surface. Air or nitrogen gas is bubbled through this mixture and the hydrophobic particles are carried along with the nonpolar gas. The copper bearing particles rise to the top of the tank with the bubbles (float) and the other particles sink to the bottom. This process is often referred to as froth flotation because of the frothing caused by the gas forced through the solution. The copper-rich material is collected and dried. The material contains about 25%–35% copper by mass and can be conveniently and economically shipped away for smelting, if desired.

Smelting is a general term for a process in which an ore is oxidized at a high temperature using gaseous oxygen. Smelting of copper ores usually consists of two stages: matte smelting and converting.

*Matte smelting* can be considered a preliminary oxidation phase in which the Cu–Fe–S ore is subjected to an oxidizing atmosphere at temperatures on the order of 1200°C. The amount of oxygen present is controlled carefully so a Cu–Fe–S "matte" richer in copper than the starting material (40%–70%) is obtained. The chemistry that occurs in the matte furnace is complex but can be represented by these unbalanced equations:

$$CuFeS_2 + O_2 \rightarrow Cu_2S + FeO + FeS + SO_2$$
$$FeS_2 + O_2 \rightarrow FeO + SO_2$$

$$FeO + SiO_2 \rightarrow FeO \cdot SiO_2 \text{ (molten slag)}$$
$$FeO + O_2 \rightarrow Fe_3O_4 \text{ (magnetite in matte and slag)}$$

All the preceding reactions are exothermic, and the process itself is *autothermal*, meaning that, once started, the reactions themselves are sufficient to keep the temperature at 1200°C.

The two layers that form, matte and slag, have different densities. The lighter slag rises in the furnace while the matte, which contains about 99.5% of the copper, settles at the bottom. The matte and slag can be drawn off independently. The matte is removed and placed in the converting furnace, whereas the slag is sent to stockpiling or copper recovery.

In the *converting stage*, the enriched Cu–Fe–S matte is oxidized to form "blister" Cu (99% pure). Oxygen-enriched air is blown into the molten matte to convert the copper to blister Cu, the iron to an iron–silicate slag, and the sulfur to $SO_2$. The converting phase is also done in two chemically and physically distinct stages. First, in the *slag formation stage*, the iron sulfides are converted into a slag material as follows:

$$2\ FeS + 3\ O_2 \rightarrow 2\ FeO + 2\ SO_2$$
$$3\ FeS + 5\ O_2 \rightarrow Fe_3O_4 + 3\ SO_2$$

This stage is finished when the matte contains less than 1% Fe.

The *coppermaking stage* results in $Cu_2S$ being converted to Cu metal and $SO_2$. The reactions occurring in this stage are

$$2\ Cu_2S + 3\ O_2 \rightarrow 2\ Cu_2O + 2\ SO_2$$
$$Cu_2S + 2\ Cu_2O \rightarrow 6Cu + SO_2$$
$$Cu_2S + O_2 \rightarrow 2Cu + SO_2$$

The product, blister copper, contains 0.001%–0.03% S and 0.1%–0.8% O. The $SO_2$ bubbles thus represented give this impure copper its name. The blister copper is poured from the converting furnace and taken to a fire refining furnace to remove the $SO_2$ and poured to make copper anodes, which are used in the final electrochemical purification step.

## HYDROMETALLURGY

Although most copper is recovered by pyrometallurgical methods, a significant and growing amount of copper is recovered with *hydrometallurgical* techniques. This approach is particularly well suited for oxidic ores or sulfidic–oxidic ore mixtures with a significant fraction of oxidic ore.

Hydrometallurgical processing of copper ores is a three-step process. First, the rock is sprayed with a dilute $H_2SO_4$ solution (known in the industry as *lixivant*) that dissolves copper and other materials from the minerals. The resulting copper-bearing leachate solution (known in the industry as *pregnant leach solution* or PLS) is purified to remove other metal ions. The purified solution serves as the electrolyte in an electrodeposition process in which the $Cu^{2+}$ ions are reduced to Cu metal.

In practice, leaching is performed by spraying sulfuric acid on a "heap" of rock piled on top of a polyethylene mat. The heap is set on an inclined slope and the solution trickles through the rock and eventually flows down the slope to a holding pool. The liquid is recycled by spraying on the heap until the copper content is high enough to send the solution on to the next step in processing.

The suitability of hydrometallurgical processing for a particular ore body depends on the nature of the minerals present. The table here presents the common copper-bearing minerals and describes how long it takes to remove a significant fraction of the copper from the rock through leaching. In general, sulfidic materials take a longer time to leach than oxidic materials.

**Typical Leach Times for Various Copper Minerals**

| SEVERAL HOURS OF AGITATED LEACHING | | SEVERAL MONTHS OF HEAP LEACHING | | YEARS OF DUMP LEACHING | |
|---|---|---|---|---|---|
| Mineral | Formula | Mineral | Formula | Mineral | Formula |
| Antlerite | $CuSO_4 \cdot 2\,Cu(OH)_2$ | Bornite | $Cu_5FeS_4$ | Chalcopyrite | $CuFeS_2$ |
| Azurite | $2\,CuCO_3 \cdot Cu(OH)_2$ | Chalococite | $Cu_2S$ | | |
| Brochantite | $CuSO_4 \cdot 3\,Cu(OH)_2$ | Covellite | $CuS$ | | |
| Chrysocolla | $CuSiO_3 \cdot 2\,H_2O$ | Cuprite | $Cu_2O$ | | |
| Malachite | $CuCO_3 \cdot Cu(OH)_2$ | Native Copper | $Cu$ | | |
| Tenorite | $CuO$ | | | | |

The presence of $Fe^{3+}$ in many ores facilitates the oxidation of sulfidic copper-containing mineral in that iron(III) can be reduced as the sulfides are oxidized.

In nature, bacteria that thrive in acidic conditions, such as those found in mines, also facilitate the extraction of copper and other metals from the minerals. Information on acid mine drainage and the role of bacteria in that process can be found at the U.S. Environmental Protection Agency (EPA) Web site: www.epa.gov.

## PROBLEMS

1. Divide the minerals listed in the first table into two categories: those that are amenable to refining via pyrometallurgical techniques and those that are not. For each of the minerals that can be refined with pyrometallurgy, calculate the amount of copper, in kilograms, that could be recovered from 1 t of the pure mineral.

2. Calculate the volume of sulfur dioxide ($SO_2$) gas at 1 atm pressure produced when 1 metric ton of ore containing 1.25% Cu in the form of chalococite is converted to native copper.

3. Using the values appended to this activity, calculate $\Delta H$ for converting chalococite to native copper.

4. A possible reaction that could occur during the matte smelting phase is

$$Cu_2O + FeS \rightarrow Cu_2S + FeO$$

a) Using the appended thermodynamic table, estimate $\Delta G$ and $K_{eq}$ for this reaction at 25°C.

b) Is this reaction likely to occur during matte smelting? Explain.

c) Do you expect the equilibrium constant to increase or decrease at the temperature used for smelting in comparison with that at 25°C? Explain.

5. Find $K_{eq}$ for these reactions, which occur during the copper-making stage of conversion in pyrometallurgy.

a) $2 Cu_2S + 3 O_2 \rightarrow 2 Cu_2O + 2 SO_2$

b) $Cu_2S + 2 Cu_2O \rightarrow 6 Cu + SO_2$

c) $Cu_2S + O_2 \rightarrow Cu + SO_2$

6. Once the copper has been extracted from the ore, it must be purified. Purification requires an incredible amount of energy. One active copper mine is at the Olympic Dam site in South Australia. Olympic Dam is an underground mine that produces multiple metals, including copper, gold, and uranium. Olympic Dam encompasses the world's largest remaining copper deposit, the fifth largest gold deposit, and the largest uranium deposit. The mine was acquired by BHP Billiton from WMC Resources in 2005.

Energy is consumed in all stages of mining. In addition to energy used for smelting and purification, electricity is also used for machinery and mining equipment. In 2004 the Olympic Dam mine used 5,477 terajoules of energy, treated 8,887 million tons of ore, and yielded 224,731 tons of purified copper. Use this information to determine the amount of energy required per kilogram of copper.

## EXPERIMENT OVERVIEW

In this activity, teams will investigate an appropriate hydrometallurgical leaching strategy for copper ore. The goal of this experiment is to leach the copper from the ore so it is present in solution. The team should design an experiment to efficiently leach the copper from ore under relatively authentic conditions. Leaching is primarily conducted using 0.1 M $H_2SO_4$, although additional solution species, such as $Fe^{3+}$ and bacterial enzymes, may be added to facilitate copper dissolution. Decisions about which reagents to use, their concentrations and volumes, and the method of mixing to employ should be made before the laboratory exercise begins. The preparatory exercises here are designed to get the group thinking about leaching strategies.

## PREPARATORY EXERCISES

7. What is the molarity of $H_2SO_4$ to be used in the leachate? Express this concentration in grams $H_2SO_4$ per liter and in kilograms $H_2SO_4$ per cubic meter.

8. What advantages might there be to using a small volume of leachate per amount of ore? A large volume?

9. Based on the options available in your lab as described by your instructor, what type of leachate mixing will be used in the study? Explain your choice.

10. Suppose your ore contains 0.20% copper by mass. What is the maximum possible concentration of $Cu^{2+}$ that could be obtained in 250 mL of leachate used to extract copper from 100 g of ore?

11. Review the information referenced in the Hydrometallurgy section on the EPA Web site. What is the optimal pH for bacteria-assisted leaching of ore? What pH is best for nonbiotic leaching? What pH range will you attempt to maintain in your study? What could occur in the mixture if the pH were to deviate beyond this range?

12. How can biological activity in the leachate be estimated?

13. Describe your leaching study. What is your hypothesis? How will your experiment be performed?

14. Suppose that a 1.00 mL sample of filtered leachate is diluted to 10.0 mL with 0.1 M $H_2SO_4$. This sample gives an absorbance for $Cu^{2+}$ of 0.335. Use the calibration plot in the following figure to calculate the concentration of $Cu^{2+}$ dissolved in the leachate.

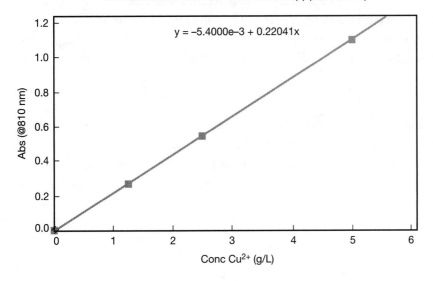

CALIBRATION PLOT FOR AQUEOUS Cu(II) (METHOD B)

$y = -5.4000e{-3} + 0.22041x$

Abs (@810 nm) vs. Conc $Cu^{2+}$ (g/L)

## PROCEDURE: LEACHING COPPER FROM ORE

Prior to arriving in the laboratory, each team will have designed an experiment to efficiently leach the copper from ore under relatively authentic conditions. Ordinarily, each team will test only one experimental leaching process, but it is recommended that all teams run an additional control experiment in which a second sample of the same ore is leached without special additives or effects. The leaching experiments will be monitored quantitatively as described here. Each leaching experiment will use 100 g of copper ore.

Once the experiment has begun, team members should divide tasks evenly to monitor the progress of the leaching experiments regularly during the laboratory period. This activity will be continued on a less regular basis by arrangement with the instructor. Depending on the resources available in the lab, options for unattended mixing methods may include shaker baths, a gravity-feed drip column, a roller mill, or occasional swirling by hand. A total leachate volume of about 250 mL is generally adequate, but teams may choose to use more or less if some advantage might be gained by doing so. For quantitative accuracy, it is important that water lost to evaporation during the experiment be replaced prior to each analysis. It is helpful in this regard to mark the initial volume of leachate on the side of the flask at the start of the experiment.

The following analyses should be used to monitor the progress of the leaching experiment.

- *Measure pH*

  A glass electrode calibrated in the acid range should be used to record the pH of the leachate at regular intervals. The object of this measurement is to allow the operator to maintain the pH of the leachate in the desired range for optimal leaching. To avoid sample contamination, it is important that the electrode bulb be rinsed free of any buffer or soaking solutions before putting it into the leachate.

- *Measure dissolved copper*

  This analysis is to evaluate the rate and overall efficiency of copper extraction in the leaching process. Prepare standards and measure the copper concentration as the instructor advises. Before collecting a sample of leachate for analysis, be sure that the leach mixture has not changed volume significantly. If significant evaporation has taken place, add deionized water and stir well to bring the mixture to its initial volume. The sample is prepared by withdrawing 5–7 mL of leachate from the mixture and either filtering (syringe filter) or centrifuging it to separate any solids from the solution. Measure the absorbance of the filtered leachate. If you use a nondestructive measurement technique (e.g., solution spectroscopy without dilution), the liquid sample may be returned to the leachate container when finished. Measure the absorbance of a standard solution or two to verify previous spectrometer calibrations.

The results of the leaching experiments should be organized to illustrate the time-dependent release of copper from the ore. The use of graphs and data tables to present these results is important. The ultimate results of the process, such as total amount of copper leached, the overall rate of leaching, and the properties or hazards of the waste products, should be summarized.

## EXERCISES

15. When conducting heap leaching, the crushed ore is sprinkled with dilute sulfuric acid, known in the industry as lixiviant. The initial lixiviant solution has the typical composition given here. The solution is recycled and therefore contains some copper prior to being applied to the heap.

a) For each species listed, determine the concentration in molarity.

| Substance | Concentration ($kg/m^3$) | Concentration (M) |
|---|---|---|
| $H_2SO_4$ | 15 | |
| $Cu^{2+}$ | 0.1 | |
| $Fe^{2+}$ | 0.25 | |
| $Fe^{3+}$ | 2.25 | |

b) Pregnant leach solution contains 8–10 $kg/m^3$ $H_2SO_4$ and 1.5–2 $kg/m^3$ $Cu^{2+}$. What are these concentrations in units of molarity?

16. The chemical reactions associated with leaching are described in words here. Write the net ionic chemical reaction that is described.

a) Nonsulfide copper ores are easily leached by sulfuric acid. For example, write a net ionic reaction for the leaching of tenorite by $H_2SO_4$.

b) An ore body contains many different minerals. When $H_2SO_4$ acts on heaped ore that contains $FeS_2$ (pyrite), the sulfur is oxidized by oxygen from the air.

c) In the presence of atmospheric oxygen, aqueous $Fe^{2+}$ is readily oxidized to $Fe^{3+}$.

d) The $Fe^{3+}$ thus generated can act as an oxidizing agent to oxidize secondary copper sulfide minerals. Use CuS as an example.

17. To get an estimate of the size of a heap, compare the following information about a heap in Arizona to the size of your school's science building. A *lift* is a pile of rock that is leached for 1–4 months. After that time, another lift is added on top of the original lift and that new ore is leached. While the new lift is being leached, the old lift continues to see lixivant solution after it trickles through the new lift.

| | |
|---|---|
| Lift height | 3 m |
| Maximum number of lifts in a heap | 25 |
| Area per heap | 12,000 m$^2$ |
| Total area under leach | $2 \times 10^5$ m$^2$ |

a) How does the area of the heap compare with the "footprint" your science building makes on your campus?

b) How high is a fully developed heap?

c) How does this three-dimensional heap compare with the three-dimensional building?

18. Using the data provided on the costs of extracting copper at the following Web sites, estimate the total cost per kilogram to produce copper metal.

Morenci Mine, Arizona:

2008 Annual Report: www.fcx.com/operations/USA_Arizona_Morenci.htm

Photos: travel.webshots.com/album/549321978hDRMRE

Economic Impact: www.istockanalyst.com/article/viewiStockNews/articleid/2942848

www.azcentral.com/arizonarepublic/business/articles/2009/01/10/20090110biz-freeport0110.html

EPA Technical Resource Document: Extraction and Beneficiation of Ores and Minerals: Copper: www.epa.gov/osw/nonhaz/industrial/special/mining/techdocs/copper.htm

Cu Mining–Producing Company: information concerning mining costs: www.ame.com.au/copper.htm

Current metal prices on world market: www.pleo.com/bcwire/metal_1.htm

Zaldívar Copper Mine, Chile: heap leaching, solvent extraction, electro-winning: www.mining-technology.com/projects/zaldivar/

## BRINGING IT ALL TOGETHER

19. On a separate sheet of paper, answer the activity question: *How is copper extracted from ore?* Provide a summary of the types of copper-containing minerals that are amenable to hydrometallurgy and pyrometallurgy. Based on the analysis of your ore, what is the preferred method? Explain your reasoning.

20. The leaching study is an open-ended project that can be designed in many different ways. Explain *your* leaching study. What was your hypothesis? How was the experiment performed? What did you learn from your study?

## APPENDIX: THERMODYNAMIC DATA

| Compound | $\Delta H$ (kJ (mole$^{-1}$)) | $\Delta G$ (kJ (mole$^{-1}$)) |
|---|---|---|
| $Cu_2O$ | −168.62 | −146.86 |
| $CuO$ | −156.48 | −128.45 |
| $Cu_2S$ | −79.50 | −86.19 |
| $CuS$ | −53.14 | −53.60 |
| $CuFe$ $S_2$ | −176.98 | −178.66 |
| $Cu$ | 0.00 | 0.00 |
| $FeS_2$ | −174.89 | −163.18 |
| $FeS$ | −100.00 | −100.42 |
| $FeO$ | −271.96 | −251.46 |
| $Fe_2O_3$ | −825.50 | −743.08 |
| $Fe_3O_4$ | −1117.55 | −1015.41 |
| $O_2$ | 0.00 | 0.00 |

| Compound | $\Delta H$ [kJ (mole$^{-1}$)] | $\Delta G$ [kJ (mole$^{-1}$)] |
|---|---|---|
| SiO$_2$ | −909.18 | −854.79 |
| SO$_2$ | −296.81 | −299.83 |
| OH$^-$ | −229.99 | −157.29 |
| Cu$^+$ | 50.00 | 71.67 |
| Cu$^{2+}$ | 65.48 | 64.77 |
| Fe$^{2+}$ | −89.12 | −78.87 |
| Fe$^{3+}$ | −48.53 | −4.69 |

Values taken from Faure G. *Principles and Applications of Geochemistry*, 2nd ed.; Prentice Hall: Upper Saddle River, NJ, 1998.

Adapted by Mary Walczak from the ChemConnections module "Should We Build a Copper Mine?" by Mary Walczak, Linda Zarzana, Doug Williams, and Paul Charlesworth.

# Laboratory: How Is Aqueous Copper Purified and Concentrated?

## LEARNING GOALS

- To explain, using Le Châtelier's principle, how aqueous copper is purified and concentrated

- To explain, using the concept of polarity, why the copper–organic complex is extracted into the organic phase

- To explain how aqueous copper is separated from other metals in solution

## INTRODUCTION

The processes by which copper ores are converted into metallic copper are complex. The two techniques typically used to purify copper are called *pyrometallurgy* and *hydrometallurgy*. The decision regarding which method to employ depends largely on the nature of the copper-containing ore. Ores with high sulfur contents (e.g., those containing chalococite, $Cu_2S$) are better suited for pyrometallurgical processing. Oxidic ores, such as tenorite (CuO), can be processed hydrometallurgically fairly easily. In addition to the role that the copper-containing minerals play in determining the processing technique, other factors must also be considered.

One consequence of pyrometallugy is that sulfur dioxide may be released into the atmosphere as part of the process. In many areas, legislatures are imposing more stringent $SO_2$ emission limits on the mining industry. Such legislation, if passed, could essentially prohibit all smelting activities for sulfide-rich ores. In such instances, hydrometallurgical processing approaches may be a viable alternative. In this activity, you will process an ore sample using hydrometallurgical methods.

Extracting saleable metals such as copper from rock using hydrometallurgical techniques involves several steps. After the rock is removed from the earth, it is crushed into smaller pieces. These pieces are *leached* (allowed to come into contact with acidic solutions to dissolve the metals present), and the copper ions dissolve in the leachate solution. The next steps in the process involve purifying and concentrating the aqueous copper solution.

*Solvent extraction* is used to purify the aqueous $Cu^{2+}$ liberated from the ore during leaching. During the leaching step other soluble ions (e.g., $Fe^{3+}$) are also dissolved into solution.

**317**

These impurity ions must be removed from solution before making the solid copper product. Solvent extraction selectively removes the $Cu^{2+}$ from the aqueous solution using a chelating agent called LIX® 984. The following graph depicts the pH dependence of the chelating agent for several metal ions. The *y*-axis is the concentration of the metal ion in the *organic* phase.

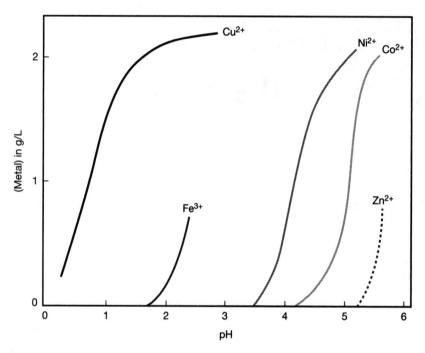

Reprinted with permission from Henkel Corporation, Minerals Industry Division.
http://biomine.skelleftea.se/html/BioMine/SX/images/pic004.jpg

Once the copper is selectively extracted into the organic phase, it can be loaded into a different aqueous phase that does not contain the contaminant metal ions that were originally present in the leachate. This solution has a different pH than that of the aqueous solution extracted by the LIX solution. This step is the reverse of the solvent extraction step and can be used to reconcentrate the copper-containing solution. This second aqueous solution is referred to as the *enriched electrolyte*.

## PREPARATORY QUESTIONS

1. Would 2 M or 0.1 M acid be more effective at extracting $Cu^{2+}$ into the organic phase from the aqueous phase? Why?

2. What are the optimal pH conditions for extracting copper *into* the LIX solution (forward extraction)? What pH is best for back-extracting copper *from* the LIX solution into a solution containing only copper ions and no other metal contaminants?

3. Extraction of copper relies on ion-exchange reactions.

   a) Write a reaction for the exchange of ions between the phases during forward extraction into the organic phase.

   b) Write a reaction for the exchange of ions during the back-extraction into the enriched electrolyte.

   c) Why is it important for this reaction to be reversible?

4. Will the pH of the aqueous phase be affected during either extraction? If so, how?

5. Explain how Le Châtelier's principle is applied in solvent extraction to concentrate $Cu^{2+}$ for electrochemical deposition of copper (also called *electrowinning*)?

6. Why is the copper complex pulled into the organic layer?

7. On a separate sheet of paper, make a flow chart illustrating the solvent extraction process in preparation for the laboratory exercise. Include typical concentrations of the $Cu^{2+}$ in the organic and aqueous phases as well as aqueous $H_2SO_4$ concentrations throughout the process.

# PROCEDURE: SOLVENT EXTRACTION

Each group will work with three different solutions:

- Leachate solution: This aqueous solution contains the copper extracted from the ore.

- Organic extractant solution: This is the LIX-containing organic solution. It is pre-loaded with a low level of copper. This is necessary to reclaim the copper removed from the leachate solution.

- Depleted "tank house" electrolyte: This is an acidic aqueous solution with about 15 g/L $Cu^{2+}$. This solution is used to collect the purified copper from the organic extractant solution.

Each team will receive 250 mL of organic extractant solution and 100 mL of depleted "tank house" electrolyte. These liquids will be used and then returned (or "recycled") to the laboratory stock for reuse as is done in commercial systems. Use these solutions to collect and enrich the copper from your leachate.

Use an analytical method provided by your instructor to monitor the $Cu^{2+}$ concentration in the aqueous phase. Either atomic absorption or visible spectroscopy are appropriate as long as the samples are quantitatively diluted to fall within the concentration range where the method is sensitive.

## Preparation of the Leachate

You will have up to three different leachate solutions. Filter the leachate with the highest $Cu^{2+}$ concentration in a Buchner vacuum funnel. Set the leachate solids aside as hazardous waste. Measure the final pH and copper absorbance of the filtrate. Be sure to consult the instructions on optimal conditions for extraction with your organic extractant before proceeding (see Questions 1–6).

## Extraction into the Organic Phase

Combine the filtered leachate (approximately 250 mL) with 250 mL of organic extractant in a 1 L separatory funnel and mix. "Mixing" of the two phases involves vigorous shaking of the contents of the separatory funnel for about three minutes interrupted by occasional venting of the funnel. Allow the layers to separate and remove a sample of the aqueous phase. Measure its pH and its copper absorbance (Method A or B). The extraction of copper from the leachate is accompanied by an exchange with $H^+$ ions from the organic phase and a decrease in pH of the aqueous leachate. For optimal recovery of copper, the pH of the aqueous phase should be adjusted as necessary to stay within the optimal range. Continue mixing the two phases in this way until the copper concentration and pH are no longer changing significantly. Remove the aqueous phase from the separatory funnel and set it aside as hazardous waste.

## Back-extraction into the Electrolyte

Measure the concentration of copper in the depleted "tank house" electrolyte as it was received (Method A or B). This is normally a solution of 2 M $H_2SO_4$ that contains about 15 g/L of $Cu^{2+}$, which represents an electrolyte from which copper has been removed in a commercial electrowinning operation. Dilute the sample appropriately so the absorbance falls within the calibrated range of the chosen method. Add 100 mL of the depleted electrolyte solution to the copper-enriched organic phase in the separatory funnel and mix. Remove a small portion of the aqueous layer and measure its copper content again. Continue mixing the two phases in this way and checking the copper concentration until the copper concentration no longer increases. Enriched electrolyte should generally contain between 20–25 g/L of $Cu^{2+}$ at this point. The used organic phase should be returned to the "solvent recycle container."

## DATA ANALYSIS

8. Find the copper concentrations of the leachate solution and the "tank house" electrolyte at the start (initial) and end (final) of the experiment.

|  | Initial [$Cu^{2+}$] | Final [$Cu^{2+}$] |
|---|---|---|
| **Leachate** |  |  |
| **"Tank House" Electrolyte** |  |  |

9. Explain how the solution is both purified and concentrated in $Cu^{2+}$.

10. Speculate on how you could obtain solid copper from your solution of purified and concentrated $Cu^{2+}$.

## BRINGING IT ALL TOGETHER

11. Answer the activity question: *How is aqueous copper purified and concentrated?*

12. Explain why the pH was adjusted to different values during the extraction of the copper into the organic layer and when it was placed back into the aqueous layer.

Adapted by Mary Walczak from the ChemConnections module "Should We Build a Copper Mine?" by Mary Walczak, Linda Zarzana, Doug Williams, and Paul Charlesworth.

# Index